I0503998

The New Age of the Automobile

And The End of Fossil Fuel

If you think you know about Electric Cars. Think again. You have much to learn

New Exciting and Affordable.
Life without gears, and exhaust fumes.
Just a powerful, virtually silent electric motor.

Electric automobiles are appearing on our roads, and the future is a new kind of driving experience. One that does not pollute the air, and promises to be smooth and delightful.
This is just the beginning. The road ahead is visible and awesome.

Self driving cars, trucks and delivery vehicles..
Batteries and re-charging whilst on the move.
Power from the Wind, Solar and Nuclear.
Generate your own electricity.
A look Beyond the Beginning.

This book covers the whole subject, and is full of information.
2018 First Edition

Author : Trevor Raven.

 Whilst a Chartered Electronics Engineer: (*MIEE, MIERE and Member, British Institute of Management*), Trevor Raven designed and built the first all electronic desktop Calculators, Digital electronic clocks, and some of the first desk-top computers. The impact of which has effected all our personal lives today. The coming of the electric car, with all the exciting technology surrounding it, feels like a repeat of that previous great social upheaval. For sure It will certainly be no less profound.

Trevor Raven's technical knowledge and experience of boardroom thinking makes him an ideal guide to describe what is available now, what is happening, and the benefits and obstacles that need to be addressed.

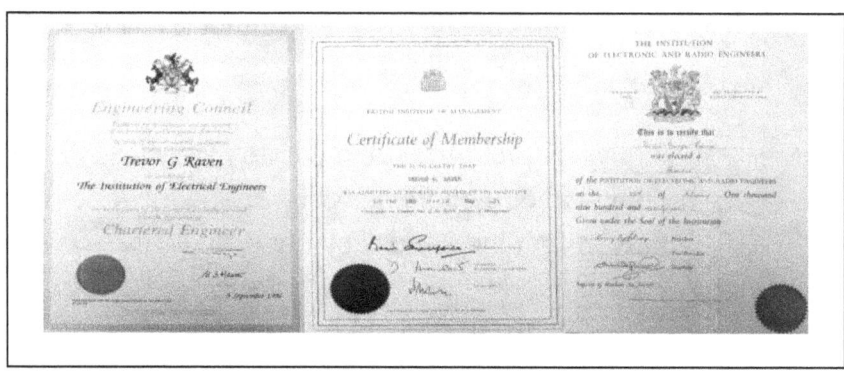

Dedicated to Daisy Wieynk

and

Ann for your support and devotion.

Dr. Malcolm Raven for your help with Maxwell's equations and matters scientific. Thanks Malc.

David Raven, Business Consultant, Author and Publisher. For your help and advice in manuscript preparation, and marketing.

It would not have happened without your input Dave.

Ken Spikings for all matters motor car.

Published by Chiltern Innovations DISS, Norfolk, IP21 4DX,

ISBN-13: 978-1717344571   **ISBN-101717344577**

Trevor Raven Tel: 01379 741285

raventrevor@gmail.com

Also by this Author

The 16 bit Microprocessor Handbook.

ISBN 0408013184

A Raven's Flight. *Published 2012.*

And

Now in preparation

Understanding Consciousness

What is it?

Is an Ant Colony conscious?

Are humans fully conscious?

The eleven parts of this book each follow a natural progression

Contents

Introduction

This book is about the coming of age of the electric car. It follows on from the biggest single event in road transport history, when in May 1885 Karl Benz in Germany, patented his Internal Combustion Engine and gave birth to the age of the motor car.

In the manuscript you will meet objects which, would quite simply, be beyond belief, to Karl Benz, and indeed, seem quite wonderful to us today. In the following pages we read why it is necessary for the responsible nations of the world to end their reliance on the fossil fuels; coal, oil and gas, and generate clean electrical energy which does not generate harmful 'Green house gas For road vehicles, this means replacing Karl Benz's wonderful internal combustion engine, powered as it is by fossil fuels, to one powered by electricity, derived from non-polluting sources. This is a world-wide transformation and will impact on every motorist on the roads after 2025. Nothing is going to stop it, and whether you like it or not, you will soon experience, an electric vehicle near you; an almost silent, pollution free vehicle with no exhaust, no gears and a straight-forward electric motor for an engine. These new motor vehicles come in every size and class, but all with similar performance characteristics to the previous generation of noisy, smoky road vehicles. Moreover, and probably of greater discomfort to the traditionalist orthodox motorist, autonomous vehicles without a driver, and aware of other road users around them will, become increasingly common by the mid 20's. We will explore the impact on Central Government finances once the use of petroleum products ceases to be used in motor vehicles. What options does the tax man have to cover the huge loss in revenue earned by liquid fuel sales?? Where will the colossal increase in clean electricity, required to power the new cars come from ? There is more than one answer, and we explore the alternatives; including, the options available to replace the huge amounts of electricity, now circulating on the National Grid, with clean power from the Wind, Solar, Hydro and Nuclear. Furthermore, how are we going to re-charge the millions of batteries which fuel these new electric cars.

The Battery.

A technical and non-technical, wide-ranging look at what is already in production, and what is imminent. This includes the wonderful world of **Graphene**, and its future role in a new super battery The power unit for these new cars is of course the battery. However, It is the one element that is not yet 100% ready, and is what prevents electric cars entering the mainstream immediately. We look in detail at the immense effort being made all over the world to produce a super battery. We discuss what can be done about the horrific casualty rates on the world's roads, and whether intelligent robot cars could be an answer. The official statistics are quite upsetting.

Re-charging the battery is another subject for which a number of solutions have emerged. We look at :- Home plug-in Chargers. Domestic, Solar and wind backup, Commercial plug-in charging bays, and Battery exchange services. There is also a fascinating introduction to Wireless induction charging, whilst the vehicle is in motion on the highway; - or in the town car park.
Our comments are clearly expressed, not only on the attributes of electric cars, but also on some negative aspects, which we feel still need addressing. For example: - Whilst the cost of an EV (*electric vehicle*) service is cheaper than a conventional car, service, who is going to repair it when something goes wrong? Not your regular motor mechanic for sure. Also, what is the

truthful, actual battery life span, before it expires. i.e. the number of charge/recharge cycles? whilst for people everywhere, the advent of HGV **Road Trains** is significant to know what is being planned; whilst intelligent cars and Electric Assisted Passenger aircraft are given a serious look, especially at the work now in progress.

Finally, we take a look at a future with batteries for all applications, that last for the whole life of the motor vehicle or other product, without needing a recharge. If you find that hard to accept, read the final chapter. "Beyond the Beginning." There is no Science Fiction in this book.

SECTION 1

The beginnings of the first Motor car age

Karl Benz

The First Electric Car

Surprisingly, the first cars in the world were Electric motors on wheels. Around **1832**, Scottish inventor Robert Anderson developed the first crude electric vehicle, powered by non-rechargeable primary cells. It could travel no more than a few yards before a change of batteries was required. It wasn't until the 1870s, that more powerful, re-chargeable batteries became available, and electric vehicles again became of interest.

In 1884 English inventor Robert Parker, using his own specially designed high-capacity rechargeable batteries, built what is claimed to be the first production electric vehicle. Pictured here in **1884** is the electric vehicle built by Robert Parker, who is driving.

Although clearly superior to Mr. Anderson's 1832 model, Parker's electric machine was still severely limited in the distance it could travel, the severe limitations of its batteries, their poor reliability and the difficulties of recharging. It was simply not suitable for anything other than perhaps a curious play thing for a wealthy gentleman.

However, at about the same time in Germany, **Mr. Karl Benz** was working on engines, and in 1879 was granted a patent on a gas fuelled 2-stroke engine. Later, in 1885, Benz designed d built his own gasoline fuelled four-stroke engine, which was then used in all his carriages.

Karl Benz and passenger
This is considered to be the world's first
production automobile.

In 1876, **Nikolaus Otto**, patented the "Otto Cycle." This is a **cycle** of engine operations which requires four strokes of the piston:-- induction, compression, ignition, and exhaust. The mixture being ignited by a spark or other means. Car makers everywhere, came to adopt The 'Otto Cycle', and still today, in the 21st century, it remains the standard 4-stroke engine firing sequence, found in all modern internal combustion motor vehicle engines.

Despite the quietness, cleanliness and simplicity of a battery powered electric engine it could never compete with the power of the **Internal combustion engine**. Its ease of refueling and, most important, the distance it could go on a tank of Gasoline, made it no contest. It would take 120 years before batteries could even start to have a chance to compete against Karl

Benze's wonderful engine. That moment could now be approaching, and the age of the battery powered electric Engine may be beginning.

Electric carriages began appearing during the 1890s, especially in France, where the firm of Panhard & Levassor were mainly responsible for making them very fashionable in Paris and The French Riviera. Today French influence is apparent in the fact that many languages have adopted the French word "automobile," not to mention, limousine, "chauffeur" and "garage"—although speakers of other tongues stopped short of calling their motor fuel "essence."

Bertha Benz

In 1872, 28 year old Karl Benz. married Burtha Ringer. She was the 23 year old daughter of Karl Friedrich Ringer a wealthy German industrialist.

At this time Karl Benz was heavily involved in developing the concept of an Internal Combustion Engine, and Burtha, who had invested half her considerable dowry into the project, and became a Business Partner, was equally committed.In May 1885 Karl Benz was awarded a patent for his new engine, but despite considerable public and international interest in this strange and wonderful **Motorwagen** there was very little inclination from the public to buy. Curiosity it seems was not enough. What was required was proactive marketing, and publicity.

Bertha realised that Karl and his company of talented

 engineers had not a clue about marketing, and the company was falling apart. So to prove that the Motorwagen had major practical benefits Bertha, on the 18th August 1888, took their two sons, and, without telling Karl', drove the newly constructed Motorwagen 106 km (66 mi) distance from Mannheim to Pforzheim, to visit her mother.

What a journey that was! Having no gas tank, she had to make constant stops for ligroin (*a mixture of hydrocarbons, and, flammable liquid, sold only in pharmacies*).

She Used her garter to repair an ignition problem, and a long, straight hatpin to clean a fuel pipe. She got a Black Smith to mend a chain, and she fixed leather onto the brake pads because they were slipping. The machine's water supply had to be filled at every stop. But in the end, Bertha reached Pforzheim. She notified her husband of her successful journey by telegram, then drove back to Mannheim next day. This had never been attempted before, and proved the feasibility of using a Motorwagen to travel long distances.

Her journey generated great public interest, and sales of the world's first motor car took off. **The rest is history**.

Today, The Bertha Benz Memorial Road is a German tourist route created to remember this attractive and very courageous lady.

By the end of the 19th century, motoring was becoming very popular, especially with the affluent members of society in both Europe and America.

In 1901 Emil Jellinek, a prominent Austrian automobile entrepreneur, joined Benze to launch the latest version of the motor wagen. Emil had a daughter called Mercedes, so the new vehicle was called the **Mercedes Benze**.

It didn't excite everyone however, and certainly not one very pompous English gentleman. The philosopher Cyril Edwin Mitchinson-Joad writing in the early years of the 20^{th,} century, clearly considers the motorist a mortal threat to all civilized, "gentle-manly" refinement. He declared:- "Motoring is one of the most contemptible soul-destroying and devitalizing pursuits that the ill-fortune of misguided humanity has ever imposed upon its credulity. The motorist is nothing more than an obnoxious show off: he desires to advertise to the world, that he has amassed sufficient wealth to hurl himself over its surface as often and as fast as it pleases him."

Clearly, Cyril was not a happy man.

How safe were these early cars?

The first cars were not particularly dangerous, although nasty accidents, *sometimes fatal*, did occur. The problem was not speed, but mechanical failure. Brakes were manual levers like those on Horse drawn carriages. They were totally unsuitable for quickly halting the motor car, even one moving at a moderate speed. Petroleum could be very dangerous and the cause of frequent explosions. Suspension and steering mechanisms were prone to fail at any time, and would never pass an inspection nowadays.

Speed Limits in the UK

Between 1865 and 1896 a motor vehicle on the highway had to be preceded by a pedestrian carrying a red flag and were subject to a speed limit of 2mph in, towns and villages and 4mph elsewhere.On 28 January  1896, Walter Arnold of East Peckham, Kent became the first person in Great Britain to be prosecuted for Travelling at approximately 8 mph, he had exceeded the 2 mph speed limit for towns, and was fined 1 shilling (5p). Mr. Arnold had been caught by a policeman who had given chase on his bicycle. In 1934 a 30mph speed limit was imposed in built up areas which remains to this day.

What about road accidents?

The **first** pedestrian to be killed by a **motor car in** the United Kingdom was Bridget Driscoll age 45 on17th August 1896)

The **first** pedestrian to be killed by a **motor car in** the United States was Mr Henry Bliss of New York state, on13th September 1899. Perhaps demonstrating that the Americans had more space than the Brits.

In Britain, 1926 was the first year that records were collected, and in that year **4,886** deaths were recorded on the roads. This shocking loss of life peaked in 1930 at 7,305. this has fallen to a low of **1,700** in 2016, or 2.9 deaths per 100,000 people.

These figures were obtained from the United Nations, World, Health Organization, (WHO) road traffic accident statistics.

In the United States, road accident deaths peaked at **53,543** in1969, falling to a low of **37,461** in 2016, or 10.6 deaths per 100,000 people.

These figures were obtained from the **"United States, Annual Road accident Statistics."**

From the above source, an additional **2.35** million people are injured or disabled every year, and over 1,600 children under 15 years of age die each year. Nearly 8,000 people are killed in crashes involving drivers ages 16-20.

Road crashes cost the U.S. economy $230.6 billion per year, or an average of $820 per person.

Road crashes are the single greatest annual cause of death of healthy U.S. citizens whilst, travelling abroad.

What about the rest of the world?

According to data sourced from the "UN World Health Organization, road traffic accident statistics:"

Road traffic accidents caused an estimated 1.25 million road deaths worldwide in the year 2010.

That is one person killed every 25 seconds.

Road traffic crashes rank as the 9th leading cause of death and account for 2.2% of all deaths globally.

The data reveals that Road crashes are the leading cause of death among young people aged 15-29. Over 1,000 people under 25 years old die in road accidents every day. Over 90% of all road fatalities are caused by human error.

Is there any solution to this catastrophe

Yes there may be; and it is from a rather surprising source.

The resolve by most nations of the world to combat the threat of climate change, and global warming, may contain the seeds of a solution to reducing the dreadful carnage on the world's roads.

At the 2016 United Nations Climate Change Conference in Paris (*more about this later*) countries agreed to take steps to try and limit the rise in average global temperature to no more than 2°C; and to do this by a drastic reduction in Greenhouse gas emissions. The unanimously agreed way of doing this is to reduce reliance on Fossil Fuels, which means changing from Coal, Oil and Gas to clean energy such as Solar, Wind, Hydro and Nuclear. It means among other things, replacing the highly polluting Internal Combustion engine with road vehicles powered by electricity. Already sales of battery powered electric cars are increasing as new high storage, fast charging batteries become available. It is the declared intention of the motor industry, to accelerate this switch, and also produce a, so called, autonomous vehicle, -- a driverless car, in the next 10 years. Already there are signs of this with more expensive models boasting automatic sat-nav control, and self parking. Near object sensing accessories are already available. Cars are becoming more intelligent, and before long will be bristling with autonomous computer controlled radar, video and proximity detectors. A self-aware computer controlled car is an intelligent car:- a robot aware of its surroundings, and programmed to obey the rules of the local Highway laws. It will not exhibit human driver characteristics of impatience, recklessness, lack of attention, and willingness to break the motoring laws. The robot's program will not allow it to do any of this, and it could even have a sensor to identify a nearby human driver, and make special allowance for his or her presence.

In conclusion Robot Cars will be very much safer machines with a huge reduction in road casualties, of which fully 90% are caused by human error. Something robots can never be accused of.

SECTION 2

Clean Air,

and the reason we have to switch to
environmentally friendy motoring

It is all about Clean Air

In mid 2017 the the UK government announced that in 2040 there will be ban on all new petrol, diesel, and hybrid-fuel vehicles from the roads. They will be replaced by none polluting, clean battery powered electric vehicles. The UK commitment, follows a similar announcement from **France**, and forms part of the 2016 Paris agreement, where 195 nations agreed to combat world climate change by reducing carbon emissions from fossil fuels, with environmentally friendly substitutes. Other nations around the world are expected to follow these actions.

The UK government's decision also comes as a result of new scientific evidence revealing rising levels of dangerous nitrogen oxide in the air, originating mainly from road vehicles. Co-author Prof Jonathan Grigg said there was solid evidence that air pollution - largely from traffic and factories, was linked to heart disease and lung problems, including asthma. As national health costs continue to escalate, with Asthma alone costing an estimated £1bn a year. "it is essential that policy makers consider the effects of long-term exposure on our children and the public purse" he said. Ecosystem scientists now believe that the growing air pollution in the UK poses the biggest environmental risk to public health, with a death rate forecast to be 40,000 lives a year, at an annual cost of some £2.7billion.

The chart below illustrates the scale of the problem. A statement from a UK government spokesman acknowledges: -- ,"Poor air quality is the biggest environmental risk to public health in the UK today, and this government is determined to take strong action in the shortest time possible".

The table below shows where these poisonous gases come

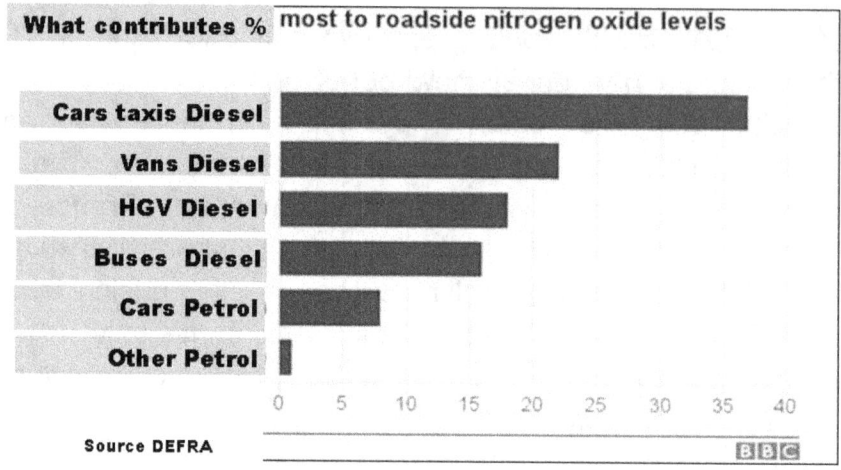

What contributes % most to roadside nitrogen oxide levels

Source DEFRA

The Paris ClimateChange Conference, 2016

Mr. Stephen Cornelius the climate change expert who represented the UK and EU at the UN climate change negotiations says that as environmental problems go, few are as big and complex as climate change. He believes it is the biggest threat faced by our world with tough social and economic issues. "It is something we can't leave to future generations to clean up." Climate change won't disappear. Rising global temperatures, which have been rising for over a century, have been speeding up and are now the highest on record. This causes melting of Arctic sea-ice, prolonged heat-waves and rising sea-levels. Most of the 195 delegates expressed their clear acceptance that current world-wide weather events are not what they have been accustomed to, and appear to be part of a world-wide change in climate. Experts add that they are not forecasting a return to previous climatic cycles. In the light of such alarming statistics all the countries agreed to work together. With binding commitments, they would attempt to reduce Global greenhouse gas emissions to below 2°C.

All those who took part in the conference recognized the following facts :-- **(a)** Climate change is a common concern of everybody. **(b)** Climate change is real and potentially irreversible. **(c)** There is an urgent need to promote universal access to sustainable energy in both developing and developed nations.

It remains to be seen how the 195 nations who attended the conference, and signed the declaration, live up to their commitments. Already in mid 2017 there is some doubt that America will confirm its commitment.

The track record of the UK

In 2015 half of Britain's electricity was generated from low carbon sources i.e. **'Renewables',** such as solar, wind and biomass (*power generated by burning wood pellets*) contributed a quarter of UK electricity between July and September 2016. The country's 15 nuclear power stations generated a further 25%. In contrast electricity from coal fired power plunged to just 3.6% of the total. This is in line with the 10 year plan to meet climate change commitments by replacing coal with the renewable sources of electricity. Wind power is playing a central role. In January 2017 The UK National Grid announced that offshore and onshore wind turbines had set a new record by generating 20% of total UK demand on a single day. Electricity generated from Solar Panels is also reporting considerable increases in electrical power from the sun and now accounts for 43.6% of the total electricity from low-carbon sources.

Mr. Peter Edwards, who developed the UK's first Wind Farm in 1991 records that after it started generating, a main criticism was that the amount of electricity contributed to the National Grid was so insignificant that he shouldn't have bothered. He now says that it is so satisfying to see just how far wind energy has come, throughout the world, and how it now competes with nuclear. As Bob Dylan once wrote, *"the answer my friend is blowing in the wind, the answer is blowing in the wind"*.

Who are the Polluters

Statistics published by the UK Department of Transport in 2017 show there are in excess of 36 million motor cars and

Top 10 Emitters

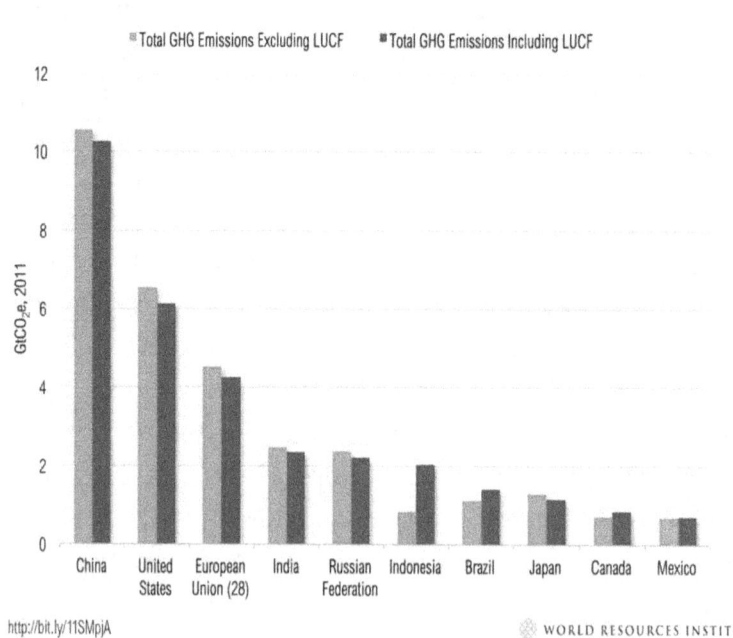

http://bit.ly/11SMpjA   WORLD RESOURCES INSTITUTE

commercial vehicles on Britain's roads, and this number of

vehicles could exceed 50 million by the switch over to clean energy in 2040. During 2017, 35.8 million tons of Diesel and Petrol fuels were consumed by road transport, and this translates to **416 Giga Watts,** which is the amount equivalent of electrical energy. It is the extra amount of electricity which must be generated from low carbon sources in order to replace the fossil fuels. What is more, all these cars, coaches, vans and lorries are becoming increasingly more intelligent. They will soon be bristling with autonomous computer controlled - radar, video and proximity detectors, giving them a somewhat scary awareness of their surroundings-- and all of them demanding electricity. The question now being asked is, when battery powered electric motors replace the internal combustion engine for use in road transport, where will all this extra electricity come from? The UK's track record of switching to environmentally friendly fuels is good. But how is it going to generate an extra 416 Giga watts?

Long before 2040 when our coal and oil fired power stations have been phased out, Wind, Solar, Hydro and Nuclear power will be supplying the Nation's electricity. The UK's track record on this topic is good, as we have seen earlier. Some shale-gas fired plants could probably remain, but most electrical energy will be generated from environmentally friendly sources.

Some industry experts are claiming that as well as a huge expansion in Wind and Solar energy, five new 3.2 Gigawatt, nuclear power stations will have to be built, at a current cost of **over** £20 billion each, in order to satisfy the demand. Although these figures, which are simple equivalents, making no allowance for the much more efficient electric motor, which does not have as many energy consuming parts such as gears and exhausts --, they also do not allow for any growth in road transport up to the change-over date of 2040. It is also not anticipated that all vehicle owners will switch to battery power. For example, some owners of heavy goods and public service vehicles may switch their internal combustion engines over to shale gas or hydrogen powered modules rather than electric motors.

Some industry observers agree, and some disagree.

Prof David Bailey, an automotive industry expert at Aston University says:-- "I think that's more than enough time. It's sufficiently long to be taken seriously, and give the industry time to adjust". Indeed, electric vehicles (EVs) could grow more than twice as fast over the next 10 years as expected just a year ago. Prof Bailey believes that by 2040 carmakers will already have switched over to battery powered vehicles regardless of this deadline, but he does think that ministers need to do more to ease the switch-over. However, some other motoring experts are far more gloomy, and warn that plans to ban the sale of new diesel and petrol cars by 2040, in a bid to encourage people to buy electric vehicles, is a very tall order that will place an intolerable burden on the National Electricity Grid.

Government ministers reply that Britain cannot continue with petrol and diesel powered cars because of the damage they are doing to peoples' health, and on a much wider scale to the health of the planet. The government believes, there is no alternative to embracing this new technology. Nevertheless, some other respected figures in the industry have warned that the Government's plan may not be feasible due to the strain this flood of electric cars would put upon the National Grid. Even the UK Automobile Associations, (The AA) is concerned that the National Grid may not be able to supply demand during the evening rush hour, quoting the Grid's own report that peak demand for electricity by motor vehicles could add around 30 giga watts to the current evening peak, which today stands at 61GW. This is an increase of 50 per cent. What is more, other voices have warned that overnight charging could trigger a colossal surge on the Grid as motorists plug-in their machines to charge after the work day.

The UK **Road Transport Authority's** own claims that the number of Electric vehicles on Britain's roads could hit around nine million by 2030, up from 90,000 in 2017. Based on these statistics, **IF** the national demand for grid electricity during the evening period is not staggered, it is suggested that peak demand could increase by 8GW in 2030. If the peak period was spread over the night-time period, this could reduce to 3.5GW. This is still a large figure but they claim it would be possible to cope with it.

NOW the Good News

Herein lies a major opportunity to relieve some of the this burden on the National Grid by generating, and storing Solar electricity, and at the same time to slash personal motoring costs. Later we show how this can be done from a home base.

The question is: could Road Tax revenues from petrol and diesel dry up? The answer is : YES, probably, and it is almost certain that governments around the world will have to devise new ways to tax road users, in order to make up for losing significant income. In the UK fuel duty is currently levied at a flat rate of 57.95p per litre for both petrol and diesel, while VAT at 20% is then charged on both the product price, and the duty. Thus, the price of a litre of petrol is compiled as follows: - pump price +vat + 57.95p. duty +vat .In the UK, and throughout Western Europe This all totals up to around 65% of the cost paid at the pump for fuel.

During *2016* the Inland Revenue received £27.9bn from vat and fuel duties, according to **the Office for Budget Responsibility.** That's getting on for 4% of the Government's total tax revenue take.

The switch to electric cars poses a big financial problem for the Exchequer, because every time a driver switches from a petrol or diesel car to an electric vehicle, the government loses 57.95p per litre in fuel tax at every fill-up.

A similar situation in varying degrees of importance exists for the tax authorities of most industrialised countries. By contrast an electric car, whose battery is re-charged from the grid electricity will currently generate just 5p in vat for every pound spent, but If the car is charged directly from solar panels on a garage roof, the tax man under current conditions gets precisely nothing. It is clear that governments everywhere will need to find alternative sources of tax revenue in order to plug the hole left by fuel duty. This will include new technologies, some of which are already available, and others will become available, when the emerging vehicle artificial intelligence becomes standard.

To help plug this income gap, a number of strategies could be phased in, as tax income from traditional sources steadily decreases.

Some options are: -
1. Annual road tax could be increased.
2. Road pricing could also be introduced with varying charges depending on the type of road used.
3. The Treasury could increase VAT on roadside battery charging points.
4. The time of day, length of time, and location could be logged and charged for. The technology already exists to track vehicles and charge drivers per mile.

Kevin Nicholson, head of tax at PwC, said the government would need to move quickly to reverse a decline in fuel duty that is already under way. The UK Treasury's official forecaster, the 'Office for Budget Responsibility', has predicted a decline of a fifth in receipts by 2021.

However, any disincentive for motorists to produce their own electricity could impact seriously on the National Electricity Grid.

SECTION 3
ELECTRIC
CARS

The Electric Car

By definition, an electric car is an automobile that is propelled by one or more electric motors, using electrical energy stored in rechargeable batteries as fuel. By 2020 there will be over 120 different car models available, claims Michael Liebreich of 'Bloomberg New Energy'. These are great cars he says, "They will make the internal combustion equivalent look old fashioned."

It is generally accepted in motoring industry circles that by the mid 2020's electric vehicles(EV's) will be rapidly replacing the old conventional Internal Combustion engines, and analysts are saying that by 2030 we could all be driving around in electric cars. The environmental advantage of electric propulsion alone is an overwhelming reason to abandon the internal combustion engine for a pollution free alternative. In Countries such as China, which has been badly hit my air pollution, much of which is from motor vehicles in its big cities. The problem is critical, and changing to EV's is already happening. The Chinese capital, Beijing suffers from crippling air pollution with frequent suspension of flights and bus services due to heavy smog. The government has launched a determined drive to get fossil fuel powered vehicles off the road, and there are large subsidies to purchasers of electric vehicles. As a result sales increased by 60 percent, to over 402,000 vehicles, in 2017. More than in the rest of the world combined. Furthermore, a project to replace all of Beijing's 71,000 petrol powered taxis with EVs was underway in 2017, and is expected to cost $1.3 billion US dollars. China expects to have 5 million EV's on its Roads by 2020. The electric vehicles now being sold in China are mainly locally-branded models that are cheap but have a shorter range than those offered by foreign car makers. The market is propped up by the huge government subsidies as part of Beijing's policy to clear the pollution from the streets of all China's cities.

Other great cities around the world which suffer from poisonous vehicle exhaust fumes could soon follow China's example. These include New Deli, Tokyo and Mexico City, London, New York and San Francisco.

The Experience of driving an Electric car.

The environmental satisfaction of owning an electric car is clear, but the delight of driving one may come as a surprise the first time. An Electric car is lighter and significantly quieter than a conventional 'petrol or diesel car. It has a less complicated construction with fewer internal parts. For example it does not need a gear train, carburetor, spark plugs, distributor or a clutch, and there is no exhaust pipe, because there is no exhaust. It is a clean quiet, environmentally friendly vehicle, and once the new super batteries become generally available, its future is assured.

In the following pages we explore some of the ground breaking developments in battery technology conceived by groups around the world, and within the next 10 years we expect that all of the major car manufactures will have their own range of EV models, designed to suite all tastes and budgets. They will be powered by the mass produced winners of this research, and will meet the target specifications described here.

- ❖ Battery capacity. 500Km (300) miles on a single charge.
- ❖ Recharge time to 80% 10 – 20 mins.
- ❖ Charge recharge life. 1,000 cycles.
- ❖ Light weight, Low cost.
- ❖ Capable of mass production.

Here we report the opinions of motorists who have switched to an electric car(EV) and consistently report how they really do enjoy the driving experience, and never want to turn back to a petrol or diesel. Owners tell us driving an electric car gives great performance and handling. They are quick, responsive and have smooth powerful acceleration. In the Northern hemisphere, One of the most-loved features of many electric cars is the ability to remotely heat or cool the car from a mobile phone, or even from a smart car key. Whilst all your neighbours are scraping ice off their cars, you can just let the car defrost whilst you eat your breakfast and then step into a defrosted, preheated cabin. Even when they have had the car for some time, some owners still feel a bit of a novelty when out in it: perhaps because of the interest shown by other drivers. They express pride in their vehicle, with the knowledge that whilst driving, they are not contributing to poor air quality, and have, environmentally, the cleanest type of vehicle on road.

As we have already mentioned, the environmental advantages of an electric car are well documented, but the delight of driving one may come as a surprise the first time. As we observed earlier. The EV is **lighter and** significantly quieter than a conventional 'internal combustion engine. It has a less complicated construction with fewer internal parts. For example it does not need a gear train a, carburetor or a clutch and there is no exhaust pipe, because there is no exhaust: -a clean quiet, environmentally friendly vehicle.

It only takes one ride in a battery-powered car to appreciate the improved ride quality. Along with the calmness and effortlessness of the drive there is more personal space inside. It really does make all but the most luxurious vehicles feel a bit clunky and outdated. Electric cars are so addictive that trying one makes you want to own it. A 21st century experience. Many people who buy one, as the second family car, quickly end up making it the prime car. At the present time an electric car costs more than a comparable specification internal combustion engine car, although incentives from most governments offer discounts to bring the price down. Notably generous are the discounts and special entitlements from the Norwegian and UK governments. There is also the warm feeling of making a contribution to battling Climate Change. Nevertheless, the bottom line coming from a survey of an affluent group of Norwegian electric car users was that about 72 per cent of buyers are choosing an electric car for economic reasons and just 26 per cent for environmental ones. These results are likely to be repeated by similar electric car users in other countries around the globe

Tax breaks for buying an electric car

In the USA Current law says that consumers who buy or lease a vehicle with a battery capacity of at least 4 kilowatt-hours in the US, qualify for a tax credit of between $2,500 and $7,500. Gasoline-electric hybrids that don't require charging are not eligible.19 Dec 2017.

The credits are available on the first 200,000 electric vehicles a manufacturer sells. Once a company reaches that threshold, the $7,500 income-tax credit continues through the end of that calendar quarter, then it is reduced by 50% every six months until it phases out completely.

Tesla is closest to hitting the 200,000 soft cap, with more than 127,000 Model S and Model X EVs, plus a relative handful of Model 3s sold, according to IHS inc. GM and Nissan are also past 100,000.

Tax incentives for buying an electric car in the UK Electric cars enjoy a generous discount of 35% of their purchase price, up to a maximum of £4,500. To be eligible, a new car must emit less than 50g/km of CO_2, and must be capable of travelling at least 70 miles in between charges. They also have to come with a warranty covering their batteries for at least three years, or 60,000 miles. Reports say suppliers seem to be meeting these conditions.

Car dealerships and manufacturers themselves can get the grants to reduce the price you pay for a brand new electric or hybrid vehicle. You don't need to do anything if you want to buy one of these vehicles - the dealer will include the value of the grant in the vehicle's price. The huge increase in electric cars in 2017 has come about because of a greater level of choice for drivers, a shift in the public's attitude towards electric cars and a constantly improving public recharging network. This means that UK electric car buyers have a greater selection of vehicles to choose from than ever before.

NOTE: The UK government's 'Plug-in' Car Grant is now guaranteed for 2018. Making an electric car a viable option for a large number of motorists thinking of changing their car.

Norway

According to a report in The Financial Times of June 14, 2017, **Norway** has become the world leader in electric cars, per head of population. Here, 35 per cent of all new cars being sold are electric. For a country whose wealth is based on fossil fuels, from deep under the North Sea this is a remarkable achievement. The country has a target of zero emissions for all new cars by 2025, and carmakers from around the world are viewing its success as a vision of tomorrow's car market. However, as electric car sales in this country of 5.2m people move from the early adopter phase into the mass market, problems are emerging for policymakers. One of the most pressing is when, and how, to rein in the extremely generous subsidies that have underpinned the boom. Adjustments so that electric cars compete with petrol/diesel vehicles, on a level playing are inevitable. Norway's Conservative government believes that the falling cost of batteries and rising demand means the moment could come as soon as 2025, while carmakers talk about 2030. "What we have proven in Norway is that if you give enough subsidies and impose enough restrictions on fossil fuel vehicles, people will buy electric," says Andreas Halse, an environmental spokesman in Oslo. He adds: "If we want to continue to be an example for the rest of the world we need to show how this can become an entirely commercial product. We need to get there because we can't rely on public finances forever." A look at some comments from electric vehicle drivers around Oslo clearly points to how accurate this assessment is.

No toll charges on the motorway. Free battery charging at charge points. No VAT or the high purchase taxes of petrol result in cutting the cost of motoring roughly in half. A general sentiment was *"To be honest, the reason for buying this car was a little bit about the environment, but mostly about the savings,"*

The bottom line coming from these affluent Norwegian electric car users was, about 72 per cent of buyers are choosing an electric car for economic reasons and just 26 per cent for environmental ones.

Countries who have announced they intend to phase out gasoline and diesel cars.

China, the world's largest car market, is working on a plan to ban the production and sale of vehicles powered only by fossil fuels.

Britain: The U.K. said in July that it would ban sales of new gasoline and diesel cars starting in 2040 as part of a bid to clean up the country's air.

Germany: Chancellor Mrs Angela Merkel, was asked if it would make sense for Germany to set a deadline to end sales of cars fitted only with petrol or diesel engines. Her reply was; "I cannot name an exact date yet, but the approach is right, because if we quickly invest in more charging infrastructure and technology for electric cars, a general changeover will be structurally possible, and Germany could ban petrol and diesel cars"

Some German towns and cities have already planned to introduce their own bans in order to curb urban pollution.

France: The government says that it wants to end sales of petrol and diesel-powered vehicles by 2040, as it fights global warming. After that date, automakers will only be allowed to sell cars that run on electricity or other cleaner power. Hybrid cars will be permitted.

Nicolas Hulot, the government official in charge of France's ecological transition, said "the goal would help the nation's automakers "innovate and become market leaders." The share of cars powered by electric, hybrid and alternative fuels in France is small about 4% but growing fast. Sales of those vehicles were up 25% in the first quarter of 2017.

India: The government said earlier this year that every vehicle sold in the country should be powered by electricity by 2030.

However, this is a desired target, said government energy adviser Anil Kumar Jain.

India suffers from acute air quality problems, having many of the world's most polluted cities. It is also a country where car ownership is increasing rapidly, as motor vehicles become more affordable to the middle classes, expected to number **547 million** citizens by 2025. If families purchasing a car for the first time buy electric, India could leapfrog ahead of many developed economies.

At least eight other countries have announced a switch to electric cars in the mid 2020's, to replace fossil fuels, according to the International Energy Agency Austria, Denmark, Ireland, Japan, the Netherlands, Portugal, Korea and Spain have set official targets for electric car sales.

The United States doesn't have a federal policy, but at least eight states have already set their own goals.

Globally, 95% of electric cars are sold in only 10 countries: China, the U.S., Japan, Canada, Norway, the U.K., France, Germany, the Netherlands and Sweden

The future looks good for plug-in vehicles. Finally, gone are the early stereotypical curiosities of slow and impractical contraptions with silent engines that ran out of fuel twice a day,-- unless they were anchored next to a power supply. Discounts and incentives from Governments everywhere are trying to persuade their motorists to switch to enjoyable, and environmentally friendly electric vehicles.

Routine Maintenance of an electric car, will, optimistically, cost about one-third the current cost of maintaining a gasoline- according to some estimates. The bottom line is that electric engines require fewer maintenance checks than petrol or diesel engines

Each year interesting new models are exhibited, and issues of concern such as battery life cost, charging time and range seem to be melting away.

Concerns about going Electric

Unfortunately , there are still a few issues, waiting to be addressed that cause unease to some astute potential buyers. These are legitimate concerns, and cannot be airbrushed out of manufacturers' literature.
They are:--

(1) What is the true battery lifetime cycle rate?
(2) Is an electric vehicle robust enough for all terrains?
(3) What about Maintenance support after the
        guarantee period has expired?
(4) How environmentally friendly is an electric car?

Here are the best answers we can get for these issues.

Issue 1. The battery lifetime charge /discharge cycle rate is, the period when the battery begins to wear out. At the heart of every electric road vehicle is of course the, battery and here, some would say, is its Achilles Heel. The battery life is limited to the number of charge/Discharge cycles it is subjected to before it is exhausted, and must be replaced. For example a lithium-ion battery should be able to perform 1,000 cycles, but this number has been shown to be critically dependent on the way the vehicle is driven. The temperature and charging - discharging current are critical parameters, that impact on the performance of the battery. They are too often specified as a brief and wooly statistic.

An important Chinese Study of Cycle-Life Prediction for Lithium-Ion, Batteries in Electric Vehicles was Published in the 'Journal of Electrochemical Science. (*www.electrochemsci.org*) It is a valuable contribution to this subject. The Authors: are Minghui Hu , Jianwen Wang, Chunyun Fu,

Here is a summary of the results of their research:-

As revealed by the battery charge-discharge cycle life tests, the fading rate of the lithium-ion battery has a non-linear relationship with the number of cycles. It is observed that the temperature is an important factor affecting the battery capacity fading rate (the higher the ambient temperature is, the greater the fading rate will be). We see that the discharge current is another important factor which influences the battery capacity fading rate (the greater the current is, the faster the battery capacity fades). It is also observed that the influence of the charge current is much higher than that of the discharge current. We know from the battery charge-discharge principle that the battery charging process is an exothermic reaction and the discharging process is an endothermic reaction. A great amount of heat is released from the inside of the battery when charging, and the battery internal temperature rises quickly due to ineffective heat dissipation caused by the restricted space of the battery shell. This phenomenon results in serious battery cathode active material pulverization and fast battery lifetime fading.

Issue 2.

How Stalwart is it ? Jason Hall, at Motely Fool inc, makes some strong points in his look at just how hardy are Electric vehicles. Consider the Land Rover Camel. The amazing pictures and video footage of these off-road vehicles has brought a  profound respect for that brand from many people. They really do seem to be able to go almost anywhere, and can handle just about everything that confronts them, whilst jerry cans solve the problem of fuel. Jason Hall believes that a tough electric vehicle has not yet been developed to cope with a harsh environment, and none have yet attempted to cross remote and hostile terrain in the way that the Land Rover Camel can. "We do not expect that to happen too soon, if ever" It will take many years for electric cars to attain a comfortable range in harsh and remote conditions on a single charge. It would have to give someone the highest level of trust in the machine's ability, without becoming a very expensive forest decoration". In other words, electric cars will not bring the complete death of the conventional vehicle for many years yet. Not in remote areas, and not in places where the climate is unfriendly. The same applies to Hydrogen Fuel Cell Vehicles, which also require a complex fuelling system.

Sorry ! but EV's are not yet ready for off-road adventures.

Issue 3. What about Maintenance support ?

For an electric car, according to some estimates, the cost of routine maintenance is about one-third the current cost of maintaining a conventional petrol or diesel vehicle. The bottom line is that electric engines require fewer maintenance checks than petrol or diesel engines That is, until the battery begins to wear out.

Despite there being no gears, no exhaust, and far fewer moving parts, electric vehicles will not bring any easy fixes for their owners. Except for a flat tire or a burnt light bulb, self repair is not an option. DIY is a thing of the past with electric cars. Even a jump start is not an option. Fortunately, the driving mechanism is a straight-forward electric motor whose rotor is connected to the car wheels. While this means that electric vehicles are inherently more reliable than conventional ones, it does raise the question of what will happen to an unlucky owner of an electric vehicle that is out of warranty and has a significant malfunction. With a conventional car, faults can be fixed in most service workshops by a skilled motor mechanic, at a reasonable charge. No such independent car service workshops at present are known to exists, anywhere. It is a very separate skill to be able to fix the electronic controls of an electric car, let alone have the diagnostic tools necessary. Hence, instead of getting your EV fixed at your local workshop, the best they can do for the motorist is help get the vehicle back to the main agent. .

Issue 4. Just how environmentally friendly is an electric car?

Although an electric car does not produce greenhouse gas emissions, the source of its power can originate from burning fossil fuels: i.e. coal, oil or gas. All of these fuels are the main source of greenhouse gas responsible for global warming and air pollution. Governments in most countries are focusing on obtaining safe, green energy from the Sun (Solar electricity), Wind energy, nuclear fission, and Hydro-electricity from waterfalls, waves and tides. Progress is positive, but rather slow, and most electricity is still being generated by fossil fuels. Road transport is a large contributor of air pollution. In the US vehicles are responsible for generating 26% of greenhouse gases, and other advanced economies in Europe and Asia will have similar figures. Charging an electric car battery will not make much of a contribution if the electrical energy stored in it came from a coal, gas or oil fired power station. In fact it will make no contribution at all. However, renewable energy sources of electricity, of the kind listed above, are increasingly coming on stream, and then, the electric car takes up its role as a major contributor to cleaning up the polluted atmosphere of our planet. At this point, you may be thinking of hydrogen fuel as the best way forward: - **Don't** ! because it generates methane, and large amounts of deadly poisonous carbon monoxide gas.

Range Anxiety' still afflicts many potential EV buyers. It stands for the worry of being left stranded with a flat battery, and no charging point available anywhere. The good news is that this situation is fully recognised and is improving fast. With Government finance charging points are being installed at  thousands of locations throughout the countryside, and this is in progress throughout the world. There are now in excess of 5,000 charging points in the UK. alone, with a website **ZapMap.com** pinpointing the nearest charging point- most equipped with rapid chargers'.

 It is reasonable to assume that before 2030, there will be a major transformation in battery technology. Already, reliable, affordable, fast charging, and safer, batteries with a claimed better than one thousand times recharge cycle, and 500 Km driving range, are being demonstrated in science laboratories. This will give an electric vehicle the same characteristics as a petrol or diesel engine. It is now just a matter of tooling up to mass produce batteries by the million. Range Anxiety will soon be a worry of the past. We think cars will come with two super-batteries, both clipped side by side into a cradle and located in a separate compartment. One main active battery, and another on standby. These two will provide a range of at least 1,000 Km, (600 miles),

Reassurance surely,for even the most cautious buyer.

New Electric Cars are appearing all the time, and sales are rising rapidly, as specifications improve, and prices come down. More than 100,000 electric cars appeared on the UK roads in 2017, compared with just 3,500 in 2013. This huge increase is occurring because of the pleasure in driving an electric car, whilst 'range anxiety, is melting away as thousands of charging points are installed. It has also got something to do with a generous contribution towards the cost by UK government to purchasers of an all-electric vehicle. During 2018, we expect to see more powerful batteries become available. Like France and Britain, countries around the world are planning to phase out gasoline and diesel cars. Norway, a country whose wealth is based on fossil fuels, has become the world leader in electric cars, as a percentage of population. Over 35% of All new car sales are electric, and the country has a target of zero emissions for all new cars by 2025. Carmakers are increasingly viewing Norway as a model of tomorrow's car market. However, as purchases of electric cars in this country of 5.2m people move from the early adopter phase into the mass market, the large incentives given are creating problems for government policymakers. Now the most pressing issue is when, and how to rein in the extremely generous subsidies that have underpinned the boom, so that electric cars will compete with petrol vehicles on a level playing field. Norway's Conservative government believes that the falling cost of batteries and rising demand mean this moment could come as soon as 2025,

Charging Bays are appearing all over Europe.

These are in Italy

Eleven
Electric Cars

Volvo will introduce its first all-electric car in 2019. The vehicle will have 250-miles of range and be offered with a range of battery options.

17 NOVEMBER 2017,

Mr. Elon Musk, the CE of electric car maker Tesla, unveiled a new model, The **Tesla Roadster**. A luxury car claimed as the "fastest production car ever" with 0-60 mph in.9secs. and 100 mph in 4.2 secs. The Roadster will have a range of 620 miles . Production is expected in 2020 at a price of $200,000

The VW eGolf
Range: 186 miles. Battery charge time 7 hours.
Power charge= 80% 30mins.
List price from US$ 30,000.   UK £ 19,320.

1

The Kia Soul.
Top speed 190 mph. Battery Charging time 4.5 hours.
Range 93 miles.

Ford Focus Electric

Top speed 84 mph. Battery Charge time 5.5 hours.
Range 115 miles.

Hyundai'Ioniq
Top speed 115mph.
Battery Charging time 4 hours. Range 155 miles.
Price: £20,156.

Renault Zoe

costs £30,000 with a range of only 107 miles on a 30kWh battery. With a home charging point it takes only 4 hours

Rimac Concept O

Car Maker   Automobilli of Croatia
Range 330 Km (210 miles)
Speed:0-124 mph in 6 secs

<u>The Fisker Emotion</u>.

has a range of 400 miles on a battery charge. It can accelerate
to 125 mph in 9 seconds and has a state of the art" graphene
battery with the "world's highest energy density" - 2.5 times
that of lithium. It has a top speed of 160 mph.
It goes on sale in in 2019, priced at $129,900.

Jaguar I-Pace All electric Sports model.
0 – 60mph in 4.5 seconds.
Range 298 miles. 30 minute recharge gives
80 mile range.
Price £63,495. Delivery July 2018.

Nissan Leaf

costs from £14,245 and comes with built-in SATNAV,
and a wall fitted home recharge box offering a full
charge in 13 hours.

Commercial Vehicles—Trucks

Buyers no longer face a lack of choice when choosing an electric vehicle — unless, that is they want a truck. Although sales of electric vehicles are soaring, car manufacturers have yet to put an electric truck on the road. This could be about to change, and it has big implications. For example, one in every six vehicles bought in the United States is a pickup truck. Ford's F-series truck has been America's best-selling pickup truck for 35 years. The company says it plans to roll out a hybrid F-150 pickup by 2020, in July 2016 Tesla's chief executive Elon Musk announced his intention to launch a new kind of pickup truck, as part of his master plan for the company. It would be unveiled in 2019. He's not the only one to get in the game, with several other American truck makers announcing plans to launch an electric pickup truck in the next few years. American manufacturing company, Workhorse Group, based in Ohio, is planning to launch its W-15 electric truck in late 2018, initially selling to fleet operators. Havelaar of Canada says its electric pickup, the Bison, will come onto the market in 2021. Via Motors Inc of, Utah is anticipates putting an electric pickup on the road in the next two years.

"Running on electricity is perfect for pickups. It makes the truck a more complete work station where you could plug in electric power tools and equipment.

Hybrid electric/gasoline engines

The Hydrogen Fuel Cell What is a Fuel Cell ?

Hydrogen vehicles run on hydrogen gas. The gas is stored compressed in high pressure tanks, located either in the rear of the vehicle or under its floor. The gas is very flammable, but it is not burned. Instead, the gas is passed to many cells, called a Stack. Each individual fuel cell has two electrodes called, respectively, the Anode and the Cathode. Electric current flows between the anode and cathode continuously when the Hydrogen gas and atmospheric oxygen are mixed together. The electricity so generated is then used to power one or more electric motors, which drive the wheels of a vehicle. The only thing emitted out of the exhaust pipe is residual water, said to be pure enough to drink. Just like electric-only vehicles, a hydrogen vehicle also utilizes a battery pack. However, since it is only used to boost acceleration and not for primary propulsion, the battery can be much smaller. It is recharged through regenerative braking and with the excess energy created by the hydrogen fuel cells. Like all electric vehicles the motor has a very strong turning torque at zero RPM, even for Heavy Goods Vehicles.

The Toyota Prius is a good example of a hybrid electric car. It was developed in 1997, and is still being manufactured by the company.

Based on smog forming emissions -The United States Environmental Protection Agency and the California Air Resources Board, both rate the Prius as among the cleanest vehicles sold in the United States

Toyota Prius HYBRID ELECTRIC

Here's what a fuel cell cluster actually looks like. This unit can produce 5 kilowatts (5000 watts) of courtesy of Warren Gretz US Department of Energy/National Renewable Energy Laboratory

The Problem with Hydrogen

When hydrogen gas is chilled to -253C° it can be stored as a liquid under great pressure in a strong sealed tank in the vehicle. When this liquid hydrogen is released it mixes with atmospheric oxygen in the air, creating an electric current flow. This drives an electric motor which powers the vehicle. A bi-product of this reaction is clean (*even drinkable*) water.

$$2\,H_2 + O_2 \rightarrow 2\,H_2O\ .$$

One often hears that powering vehicles with hydrogen fuel cells is environmentally friendly, and that the only emission is water. Whilst it is true, that vehicles powered by liquid hydrogen gas and oxygen from the air do not produce greenhouse gases, just water (*see above*), in their exhaust. It is also true that the most abundant element in the universe is hydrogen, virtually all of which is bound up with other elements. For example 2/3rds of all water is hydrogen. It is the manufacturing process in extracting the hydrogen gas from the ground or from water, which generates large amounts of Green House gas, and damages the reputation of hydrogen as a fuel to replace petroleum. With present day extraction methods, Hydrogen processing is certainly not climate friendly.

Electric Trucks and Road Trains

The Tesla Inc. plan for new electric vehicles goes further than just trucks. The company is betting on both electrification and **autonomous driving systems** that could eventually remove the need for a driver, but in the short term could increase efficiency and safety. The company has revealed plans to test long-haul, electric trucks that move in so-called **Platoons**, or **Road-Trains**, that automatically follow a lead vehicle driven by a human. Computer-controlled trucks driving much closer together could increase the aerodynamic efficiency of the chain of vehicles. The front vehicle takes the brunt for the whole group by pushing the air out of the way as it travels along the road. This reduces drag for the rest of the platoon and therefore reduces transport costs and emissions.

The convoy idea was first tested on a public road in Spain by **Volvo** of Sweden. Now **Scania Trucks inc**, also of Sweden, is planning to test Road Trains claiming these could cut emissions by 10 percent due to the decreased drag between the closely coupled vehicles. That also means less fuel consumption. In 2017 The United Kingdom Department for Transport announced that platoons of self-driving trucks will be trialled on Britain's motorways in 2018. Tesla also has been seeking to test its electric trucks in Nevada and California. These vehicles will be electronically linked together, and driven as 'road trains. All these tests connect groups of trucks in sync with one another, snaking along motorways with just a few feet between their bumpers. The truck in front is still driven by a human, but the controls will be shared with all other vehicles in the convoy over an intelligent wi-fi loop. The group will act as one, whilst sensors on the vehicles keep the carriages in check.

Automating trucks, as opposed to smaller vehicles, seems like a no-brainer: They travel long distances, often with heavy Loads and, no doubt because of that, driver concentration can Become a real safety issue. It therefore makes sense if every vehicle in a group on the road is automated and able to communicate with each other.

The Volvo Concept Truck is the result of a Swedish and American cooperation to develop a heavy duty goods vehicle with a unique **Electric- Hybrid power train**, which recovers energy when driving downhill, or when braking. The recovered energy is stored in the vehicle's batteries and used to power the truck in electric mode whilst on flat roads or low gradients. In long haul transportation, it is estimated that the hybrid power train will allow the combustion engine to be shut off for up to 30 per cent of driving time. This will save between 5-10 per cent in fuel, depending on the terrain the vehicle is travelling along. Also, developed specially for the hybrid power train, is an enhanced version of Volvo Trucks' driver support system called 'I-See', which analyses the upcoming topography using information from the GPS and electronic maps. The first Electric trucks and Road Trains will appear with Hybrid diesel/Electric power units. Some especially in Asia will probably run on varying types of Hydrogen fuel.

Wireless Coupled Road Trains

We believe that Gradually, as battery power grows, trucks will move over to all Electric. Power Since this is becoming the most economic, and easily obtainable fuel

Robot Cars with no Driver

Despite what most people may think vehicles that drive themselves are not the stuff of science fiction, or if not, are many years away in the future. The fact is that Google, and other companies around the world, are working hard to develop self-driving cars. Cars that are aware of their surroundings, and programmed to safely control the vehicle they are part of. These automated machines on wheels have state of the art computer brains, that understand in detail the local highway driving laws, and are equipped with Sat Nav and an array of radar, laser, video, and acoustic proximity detectors, making them aware of the moving world around them. With these tools they can navigate along city streets or through the countryside at a level of safety not reached by a human driver. It can already be demonstrated that not only can a robot offer much improved safety for other road users, pedestrians or drivers, but by its awareness of traffic congestion ahead is able to choose a route with fewer traffic jams.

Indeed, if the claim that a robot, *unaided*, will be able to drive a vehicle more safely through a congested city street than the average human driver seems absurd, then consider what is already available today, in 2018, from any authorized car dealer near you. Then discover that virtually every car manufacturer is making plans for automated systems.

At the present time Google seems to be at the forefront. Its Subsidiary company, 'Waymo Inc', is working on the development of a driverless car based upon a Toyota Prius hybrid electric car, and has created considerable public interest with its self-driving test run. This is a car which cruises along a stretch of busy road in Las Vegas city. There is no driver at the wheel, but a technician is in position in the car – just in case.

Cars nowadays come with a bewildering array of driver-assist accessories, sensors, and aids. **These include**:-

- ❖ GPS navigation.
- ❖ Rear view video for parking.
- ❖ Adaptive cruise control, to slow or stop to avoid collision.
- ❖ Fuel meters, displaying the remaining range.
- ❖ Open door indicators.
- ❖ Anti-skid braking.
- ❖ Air bags throughout the car interior.
- ❖ Automatic headlights, and windshield washers.
- ❖ Ultrasonic close proximity detectors, at front sides and rear.
1. There are also accessories for automatic car parking. and Lane nudging in case you drift out of your lane.
2. And soon to come, a camera to monitor the driver's eyes and detect symptoms of drowsiness.

Although not available yet, sensors are being developed, able to monitor the driver's body condition. With the help of heart rate, pulse and temperature devices attached to the steering wheel system the software decides whether the driver's condition is safe enough to drive the car. If an onboard breathalyzer decides that the driver is over the alcohol drink limit it will not allow the engine to be turned on.

However, If the system indicates that the driver's medical condition does not allow him to safely drive the car, it can block the engine, and if necessary, gently park at the curb.
At a command from a human – who could be a passenger, the system can be enabled to telephone for assistance.

These features are additions to the sensors already listed above, but the car is not equipped to make decisions.
It is not yet autonomous.

It is important to understand that at the present time (2018), all the data collected by the above sensors is reserved for the sole attention of the driver, and of course the car occupants.

However, data sourced from similar sensors by Google and others is accumulating on an enormous scale. The company's own self-driving cars have already logged more than 250,000 miles on public roads. A massive database is growing.

Inter-vehicle communication is sure to be critical in the evolution of a truly autonomous vehicle. Sensing a wi-fi- signal from a nearby vehicle, establishes its position with reference to your own. The two vehicles can then exchange information with each other about braking, steering, speed and direction. It is then easy for the computer to calculate where the other vehicle will soon be, and so, avoid a collision. Furthermore this data can be updated ten times a second.

Other technology, known as dedicated short range communications (**DSRC**), enables a car to tell an intersection that it is approaching. A computer system at the intersection would then be able to coordinate all the approaching cars — assuming they were self-driving cars — and funnel them through the crossroads without stopping. The reduction in crashes could be dramatic.

Support for this study comes from the US National Highway Traffic Safety Administration, and concludes that various forms of vehicle-to-vehicle communications could cut all crashes by up to 79 %

Artificial Intelligence (AI)

In order to safely navigate, through towns and countryside, an autonomous car will have to be proficient in the use of a self-learning computer operating system, and be able to interrogate the computer's massive database, to decide on the appropriate action to take in any given situation. (*a number of these 'self-learning' computer programs exist*). Cars talking to each other, learning where they are going and what their actions are, is going to take some years to perfect, before truly autonomous cars can be handed over to the robot and allowed to travel freely on the roads. However, It may come as a surprise for many people to learn that this Self learning software' is a field of computer science that is now well established, and enables computers to learn without being explicitly programmed. Using machine learning algorithms to build artificial Intelligence programs (AI) and mine the huge databases now being amassed, will enable the on-board computer to decide for itself on a safe strategy to handle any road traffic situation with confidence.

It may take 15 years before fully autonomous cars can be released on to the world's roads, but it will happen, and it may happen sooner than we think. Within five years it could be perfectly feasible for the information produced by the majority of today's in-car sensors, to be uploaded to a huge 'cloud based' database, and accessed by the car's own central processor. Within 10 years Cars could appear that can drive themselves, albeit with some help and guidance from a human. Soon after that, the car will understand and take full control. The word automatic finally takes on its true meaning.

As has been suggested previously in this book, a huge benefit of self-driving road vehicles is the predicted drop in road accidents. It could be a possible solution to drastically reducing the terrible carnage on the world's roads and it rests on the effort to combat climate change. A central highlight of the International community's attempt to combat global climate change is to reduce carbon emissions into the atmosphere by phasing out greenhouse gas generating fossil fuels. This means road transport converting to electric cars and trucks, and electricity of course is perfect for computer controlled vehicles.

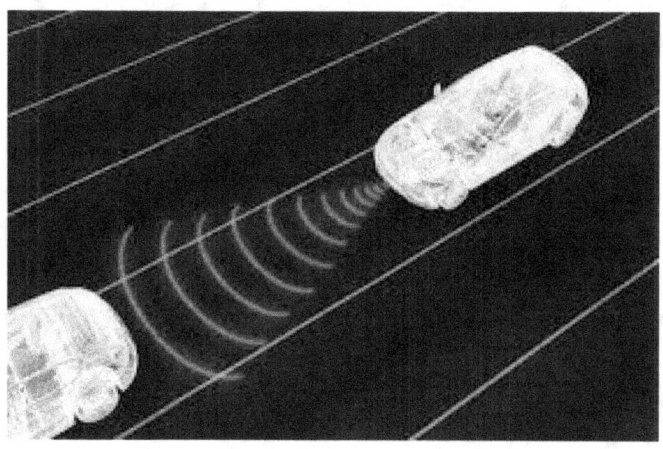

Few people would question the assertion that the most important reason for self-driving cars is to save lives, and reduce injuries. The technology could save many thousands of lives a year, in the US alone. Furthermore, it could slash car accident costs which according to the American Automobile Association cost $300 billion a year. The fact is too many humans are, quite frankly, terrible drivers. Some 90 percent of all crashes are caused by human error. A self-aware computer controlled car is an intelligent car. It is a Robot, aware of its surroundings, and able to communicate with other robot cars nearby. It is programmed to obey the local highway laws. It will not exhibit human driver characteristics of impatience, recklessness, lack of attention, and willingness to break the motoring laws. The robot's program will not allow it to do any of these things. Robot cars could even have a sensor to identify a nearby human driver, and make special allowance for his or her presence. In conclusion Robot Cars will be very much safer machines with a huge reduction in the terrible road casualties, of which fully 90% are caused by human error. Something robots can never be accused of.

Autonomous car with no driver

Googles modifiedToyota Prius uses an array of sensors no navigate public roads without a human driver.Other components not shown include GPS

LIDAR Rotating sensor on the roof, scans more than 200 feet in all directions to generate a precise three dimensional map of the car's surroundings

Position estimator sensor mounted on the left rear wheel measures small movements made by the car and helps to locate its position on the map.

Video camera, mounted near the rear view mirror, detects traffic lights and helps the car's on board computers recognize moving objects like pedestrians and cyclists.

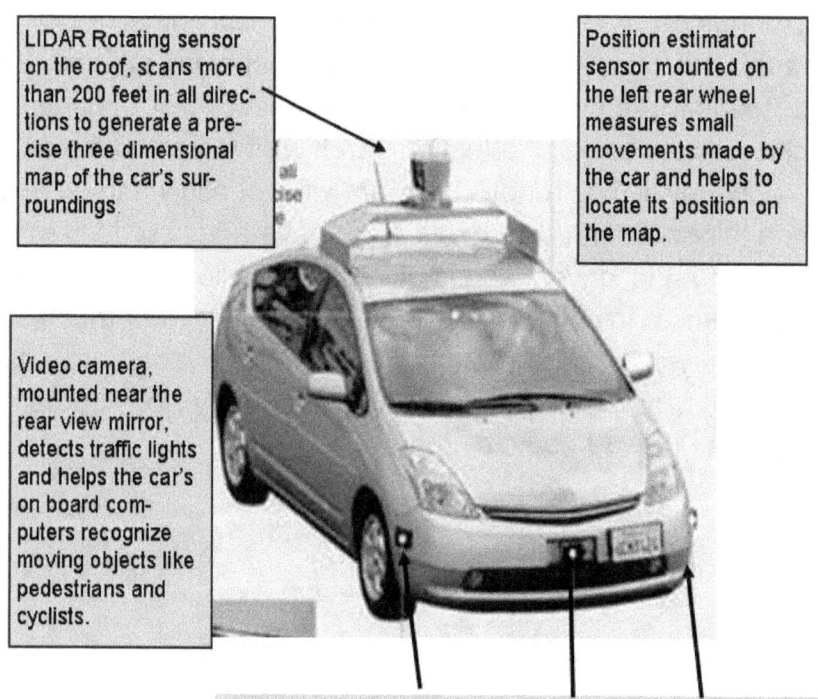

4 standard automotive radar sensors. 3 in front and 1 at rear.help determine the position of distant objects.

Drones and electric Passenger Aircraft

An electric aircraft is a flying machine powered by one or more electric motors, and is surprisingly not new.
Electricity may be supplied by a variety of methods including; batteries, and small turbine generators.
Battery powered drones have existed since the 1970s.
In the 21st century significant development has taken place with drones, becoming widely used for many applications such as surveillance, Crop analysis, Aerial photography, and an increasing number of military uses.

The first manned free flight by an electrically powered aeroplane was not made until 1973 and most manned electric aircraft today are still only experimental demonstrators. However, Between 2015 and 2016, **Solar Impulse 2** completed a circumnavigation of the Globe.

New developments are now rapidly taking place.

Zunum Aero is an aircraft manufacturer startup based in Kirkland, Washington. The company is backed by Boeing HorizonX and JetBlue Technology Ventures. They have been working since 2013 on a family of 10- to 50-seat hybrid electric regional aircraft.

UK airline Easy Jet has taken a step towards a future without jet fuel. In April 2017, amidst much publicity from an excited press, it has announced plans to participate in the development of a battery powered passenger plane. The company said it had

partnered with US aircraft designer **Wright Electric** to develop an electric aircraft for flights under two hours. If the project works, it will cut emissions, noise and fuel consumption by a significant amount. Company Chief Executive, Dame Carolyn McCall, said, "For the first time in my career I can envisage a future without jet fuel" and we are excited to be part of this project." The new, aircraft would have a range of 335 miles, and this means it would be able to fly from London to Paris, Brussels, Amsterdam, Cologne, Glasgow or Edinburgh. Easy Jet's main contribution will be to provide advice and industry expertise that will allow the US designers to develop a battery-propelled short range commercial airliner with the above operational specification. **Wright Electric** was founded in 2015 by US entrepreneur Jeffrey Engler and electric aircraft designer Chip Yates. It already has a prototype two-seater electric plane, and is working with several airlines to convert an existing 120-seat passenger plane into a hybrid plane. The project would be completed within a decade, but they are unsure at this stage how much building the first plane would cost.

Rolls Royce Aero Engines are Collaborating with Airbus and Siemens on a project called 'E-Fan X' to build a hybrid electric plane by 2020. The three  companies aim to build a demonstration model based on a BAE 146 short-haul passenger jet. They will convert it to run partially on electricity.

A spokesperson for the group said the technology could mean cleaner, quieter and cheaper journeys. They also raised the prospect of radically changing aircraft and airport design, allowing air travel to supplant rail for many more intercity journeys. The group is in talks with the British government to partially fund the joint project, which could cost hundreds of millions of pounds.

The demonstration model will have an electric unit, powered by an onboard generator to replaces one, and eventually two, of the plane's four gas turbine engines. Mark Cousins, the head of flight demonstrators at Airbus, said: "We decided we needed to be more ambitious because the world and technology is moving so fast."A number of airlines were interested, he said. "The objective is to reduce environmental impact and significantly reduce fuel burn." Paul Stein, the Rolls-Royce chief technology officer, said: "Aviation has been the last frontier in the electrification of transport, and slow to catch up. This will be a new era of aviation."Electric motors, which can be tilted more easily, could lead to radical changes in overall plane design. Quieter and cleaner travel could also mean airports can be situated much nearer city centres and, particularly in the developing world, less need to build or maintain infrastructure such as rail lines, Stein said. Frank Anton at Siemens believes the project has the potential to move mass transport from rail to air; flying might become the norm."For now, challenges include accommodating the weight of a two-tonne battery storing electricity onboard a small plane. "We have to get more than 10 times the power out of the same weight."However, Anton said he believed he would one day buy tickets for short-haul commercial flights using hybrid electric planes.The addition of Rolls -Royce to the Airbus-Siemens collaboration will step up the race to transform gas turbine aircraft.

Companies that did not Survive.

A look at six pioneering companies, set up to make Battery fuelled Electric cars, but for various reasons Did not succeed.

If there's one lesson to be learned from these companies perhaps it may be this:- **"Starting a new type of car company takes a huge amount of money, orders of magnitude more than most venture capitalists who fund Silicon Valley Software start-ups are prepared to fund. If the cash runs out at any point, the whole company will die. Not just a single product".**

The Fisker Electric Automobile

Henrik Fisker is a well known Danish-born automotive designer and entrepreneur based in California. Fisker's first electric car went up in flames, "literally". His company Fisker Automotive was the inspiration behind an electric hybrid called 'the 'The Karma. In 2011 reports published by Wire Magazine claimed that the $100,000 luxury car had a host of battery issues that resulted in 600 cars being recalled. Worst still, on more than one occasion, a Karmer auto was reported to have burst into flames. This finally culminated in the close down of the company. Henrik Fisker is highly regarded as the designer behind the iconic cars, BMW Z8 and the Aston Martin DB9. He was never likely to leave the automotive industry, and in 2017 unveiled a new supercar the '**Force 1**', This new electric car has a claimed range exceeding 400 miles on a single battery charge: something considered outlandish just 2 years ago, but not   now.
The car will be on sale in 2020. At a price of around Two Hundred Thousand dollars.

Mahindra and Mahindra Limited

This company was founded in 1945 and is an Indian multinational manufacturing corporation with, headquarters in Mumbai. It is one of the largest vehicle manufacturers by production in India and claims to be the largest manufacturer of tractors in the world. In 2015 the company announced its subsidiary, Mahindra Automotive North America inc. was planning a new range of electric cars including premium models to rival the Tesla Super car. Since then all has gone quiet, and it is now not certain if the company has abandoned, or is continuing, with this project.

Li-ion Motors Corp.

This company was incorporated in Nevada in 2000 as a 'development stage technology company' focusing its resources on the development and marketing of lithium-powered road vehicles and products. In September 2011, the still relatively unknown electric car maker won $2.5 million in the, Automotive X-Prize competition. The company then began putting out regular press information about its Prize-winning **'Wave II' electric car,** and especially, its super-hyped but never-quite ready, **'Inizio'** super-car. This supercar had a claimed top speed of 170 mph and could go 250 miles on a single battery charge. However there were frequent production delays, and the **Inizio** super-car.was never able to demonstrate its prowess. By mid July 2013 the company was experiencing serious cash flow problems, and on July 31, 2013, it reported a net loss of over $1 million for the prior twelve months. The company was dissolved in 2014

Better Place Inc

Another casualty of the electric car revolution was the highly Promising electric Car Start-Up, Better Place inc.

The company launched its electric car service in Israel in October 2007. Like wireless phone companies that discount their hardware and make money by selling minutes, Better Place sold its electric car at a

considerable discount, then leased the battery to the buyer for a monthly payment. When a new battery was required, the car was driven into one of 37 automatic switching stations where the old battery is removed and replaced with a fully charged new one -- All within five minutes, automatically, and without the driver leaving the car. Despite the innovative battery-swapping technology, and a major sales campaign, the company sold only 750 cars in two years, and with $500 million losses piling up was forced to close in July 2013.

Coda Automotive inc

The Coda was a 4r-door, electric car manufactured in China. After being re-scheduled several times, deliveries to retail customers in the United States began in March 2012. The car was sold exclusively in California but only 117 units were delivered. The experiment obviously failed, spelling the end for a unique California-China electric car collaboration, and to no surprise the company closed down in 2013.

Aptera Motors inc.

(*Aptera is a Greek for word "without wings"*) and was an American start-up company founded in 2006 with a mission to build high-efficiency electric road vehicles. The company was based in Oceanside, California. By 2008 it was suffering severe cash flow problems, and was running out of working capital. Rescue came from a Chinese company who acquired Aptera and all its Intellectual property. A company called 'Zaptera USA Inc. Was formed to build a range of electric cars. Aptera's intended first product was a three-wheeled two-seater named the Aptera2. The claimed fuel efficiency of the new car was an impressive 0.78 L/100 km. when plugged in every 120 miles. (190 km), this would make it one of the most fuel-efficient cars in the world. In 2010 Aptera2 was entered in to the **Automotive X-prize** competition but, didn't finish the 50 lap trial, as the car overheated after 30 laps.

Something was clearly wrong with the car, and in August 12, 2011, Aptera started to return deposits from customers. Finally, n December 2011 the company closed down.

SECTION 4

The Battery

Heart of the machine

It has been calculated that all the batteries on Earth can only store 10 minutes of the world's electricity needs.

The Brookings Institution

This highly respected, nonprofit public policy organization is based in Washington, DC. Its mission is to conduct in-depth research that leads to new ideas for solving problems facing society. Here is what they have to say about batteries.

Advancement in battery technology would not only transform the transportation industry, but also significantly affect global energy markets. The combination of batteries with renewable energy sources would diminish the need for oil, gas, and coal, thereby altering the foundation of many economic and political norms we currently take for granted. However, we certainly don't have to wait until the perfect battery is developed to give tangible improvements in performance. Despite the current shortcomings of today's batteries, the potential global impact that even relatively moderate improvements can have is astonishing. The challenge of creating a new type of battery for motor cars and the amount of money being invested by governments, and the motor industry itself, has triggered a frenzy of activity in established companies, startups and University groups; at a level not seen for 150 years. In this section we look at the technology and some of the companies and individuals working on new types of battery with a reasonable degree of certainty that one of these batteries will turn out to be the winner. The criteria of 500 Kilometers (300) miles on a single charge with just a few minutes to recharge, will obviously have to be robust. One great advantage of these new batteries to car designers is that they can take potentially, more or less any shape,

giving much greater design freedom than with traditional vehicle design. AS new electric car models arrive on the scene it is clear that they are finally gaining real traction in the market. However, battery performance is very much at the heart of the electric car, and it is just not there yet. The battery cost makes electric cars more expensive than their gasoline counterparts, and even though battery technology has certainly made big strides towards the target specification, much more research and development has to be done to improve performance, and trigger a mass switchover to electric vehicles.

So, what are the newest innovations in battery technology, and what do such advances mean for the electric vehicle market?

The Cinderella Product

Batteries had a bright future until the advent of Michael Faraday's Electric Generator in the early 19th century. Building a new electric motor or electric generator became a whole lot more interesting, and everyone wanted to build one. Battery development was pushed into the background. It became the Cinderella of electrical science. The development of ever more powerful electric machines surged forward. Indeed it was recognised at an early stage that although batteries were of value for experimental purposes, and also in applications requiring relatively small amounts of electricity, such as electro plating and the new electro-telegraph, they could not provide a large current for a sustained period of time in applications where a large amount of energy was required. The battery was just not up to the job. In addition, the only batteries available at the time were what we call 'Primary Cells'. They can only be used once, after which their electrodes are irreversibly changed, and they cannot be recharged. Even when a rechargeable battery emerged it was not powerful enough to replace an Electric Generator; and this situation continued right through into the 21st century. The battery hasn't advanced in decades, in sharp contrast to electronics technology. This has raced ahead in step with **'Moore's Law',** which states that, "The device density of a silicon chip doubles every two years". As a consequence, in the age of powerful pocket computers, development of battery powered mobile devices is restricted, and hundreds of millions of users around the world find themselves greatly inconvenienced by batteries that simply do not have enough power to last for more than a few hours, before a recharge is required. "We need a miniature battery that will last for weeks, and preferably, months before needing a rapid recharge; and we need a car battery powerful enough to drive a car for the same distance as a tank full of gasoline. Solar power and wind power are an excellent source of electricity, but they are both intermittent, available only when the wind is blowing or the sun shining. It is clear that a form of energy storage is required to even out the periods of availability. It follows that a very large capacity battery is needed to store energy for use during the no sun and no wind periods, and all around the world scientists are working hard to come up with this super battery. One designed to store large amounts of electrical energy, and release it steadily for applications in:- (a) Very small amounts over long periods, as in i-phones. (b) Kilo watts for medium periods as in Electric vehicles. (c) MegaWatts for short periods in Grid Downtime bridging.

Furthermore, our super battery should be capable of being recharged in minutes rather than hours.

This is a very challenging objective, but the prize is huge, and the competition to build a viable super battery is fierce. Billions of dollars are being invested by car manufacturers, and technology groups all over the world. In July 2017 The UK Government announced yet another tranche of £246 million for the 'Michael Faraday battery development initiative'. *A serious move, but strangely inappropriate for the one name innocently responsible for the battery's 150 year Cinderella status was* Michael Faraday. In addition, the UK Government is setting up a "battery institute" to award hundreds of millions of pounds for super battery research. Likewise, the US Government has invested $2.4 billion into Battery research, and the German Government recently invested another 35 million Euros. Other countries are also awarding their Universities with grants for battery research. Funds are pouring in, and things are starting to happen The 150 years of neglect for batteries is at last being undone. A new age of super batteries could be just around the corner.

To follow this subject further and in more detail, an understanding of Battery basics is probably going to be useful.

The engine which drives an electric car is basically a modified, Electric Motor, easily recognised by any electrician. However, at the very heart of the vehicle lies the battery, the most important element of the system. It is the equivalent to the fuel storage tank. It is where the driving energy is stored in a chemical state.
The Battery supplies electricity to the electric motor, which is the equivalent of the 'Internal Combustion Engine', turning the wheels of the vehicle. The battery is fundamentally, a mobile source of power, allowing the electrical to appliances to work without being directly plugged into an outlet. While many types of battery exist, the basic concept for all types remains the same; one or more electro-chemical cells converts stored chemical energy into an electric current. A battery is usually made of a metal or plastic casing, containing:-- (1) a + positive terminal (**the anode**). (2) a - negative terminal (**the cathode**).

(3) an **Electrolyte**. This is a chemical that allows **ions,** (*charged particles*) to flow between the cathode and anode. In a nutshell, the electrolyte itself is a chemical that allows the flow of electricity.

Electrolytes found in modern batteries include:-

Lead Acid. The electrolyte is sulfuric acid.
Nickel-cadmium (NiCd) The electrolyte is potassium hydroxide.
Nickel-metal-hydride (NiMH) uses the same as NiCd.
Lithium-ion (Li-ion) individual Lithium–ion phosphate.

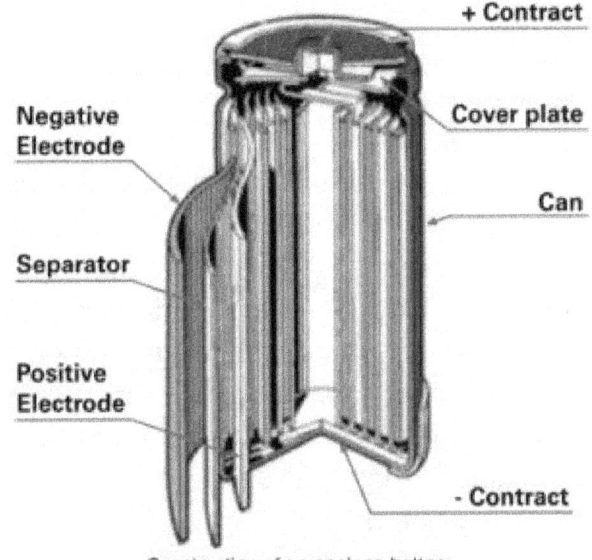

Construction of an eneloop battery

When the circuit between the two terminals is completed, the battery produces electricity through a series of reactions. The anode experiences an oxidation reaction in which two or more ions from the electrolyte combine with the anode to produce a compound, releasing electrons. At the same time, the cathode goes through a reduction reaction in which the cathode substance, ions and free electrons combine into compounds. Simply put, the anode reaction produces electrons while the reaction in the cathode absorbs them and from that process electricity is produced. The battery will continue to produce electricity until electrodes run out of necessary substance for creation of reactions.

❖ Here is a target battery specification for an Electric vehicle to replace an Internal Combustion Engine.

   Range capability : 500Km (300 miles) single charge.
❖ Recharge time 10– 20 minutes. (80% full)
❖ Charge recharge life. 1,000 cycles.
❖ Light weight.
❖ Low cost.
❖ Capable of mass production in.

So, the challenge is to build a battery that will power a medium size motor car for 500 Km on a single charge, with a recharge time of around 8 minutes, and charge-recharge cycles in the range of 1,000. As reported earlier, Competition is fierce with governments, big technology companies and car manufacturers all over the world investing resources into battery research; in what seems like a frantic effort to undo the 150 years of neglect, triggered by the discovery of the electric generator and electric motor.

.Now things are starting to happen. A new age of **super batteries** could be just around the corner. The technology is poised to overcome the curse of 'intermittency' that has long bedeviled wind and solar Power. With a super battery, electrical energy will be stored for use later at times when the sun sets, and the wind is calm. Amazing new '2-dimensional materials' such as Graphene have been discovered, and its application in batteries is now being explored.

Already there has been a plethora of high capacity battery powered products being announced, or being developed. Batteries in many shapes and sizes, from wearable battery powered clothing to miniature cells used to power hearing aids and wristwatches, right up to to battery Road Trains, Passenger Aircraft and Electric Grid size Mega storage batteries. Some of which we examine in the following pages.

Super Batteries.

Battery technology is changing fast, but still has a long way to go. In particular, today's batteries for electric road vehicles (EV's) seriously limit the range of the vehicle, and require charging times measured in hours. Indeed, Range is the main limiting factor in the uptake of EV's. High capacity batteries which are light, cheap, and with fast recharge cycles are urgently required. The good news is that real progress is being made, and Research teams around the world have begun to demonstrate enhanced performance storage devices, which promise to release the log jam. One of the new materials is Graphene. This is a major discovery in the effort to develop a super battery.

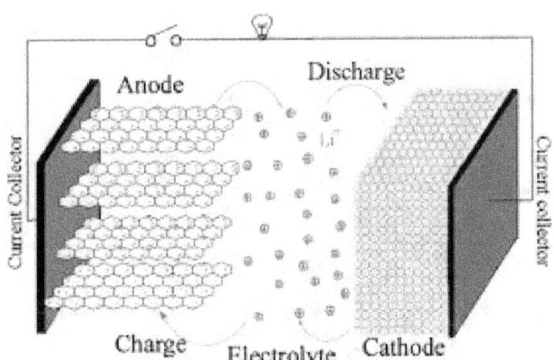

Welcome to **The Wonderful World of Graphene**

If you have ever drawn a line with a black pencil, you've probably deposited a layer of black graphite on to a piece of paper. In fact you have made graphene. Scientists knew that this is a layer of one atom thick, crystal of graphite sheets, stacked one on top of another. The problem was that no-one had worked out how to extract a layer from the graphite block. Until that was in 2004, at the University of Manchester UK, when two pioneering scientists , Andre Geim and Kostya Novoselov, isolated a single layer of graphene and were able to reveal its amazing properties. For their stunning scientific feat Professor Geim and Professor Novoselov, were awarded the Nobel Prize for Physics in 2010

Graphene explained

In a powerful microscope a layer of Graphene looks like a layer of chicken wire with a honey-comb structure, made of carbon atoms and their bonds. Since plain Graphite is many graphene sheets stacked to form a tower, then three million graphene sheets would be one millimeter thick.

The current excitement about graphene stems from its remarkable physical properties, and the potential applications these properties could offer for the future. Graphene is the world's first 2D material. It is the thinnest substance ever made, and is extremely flexible

A single sheet of carbon atoms arranged in a hexagonal honeycomb pattern is as stiff as diamond and 200 times stronger than steel. Yet at the same time it can be pliant and rubberlike. It conducts electricity faster at room temperature than any other known substance, and it can convert light waves into an electric current. In the decade since graphene was first isolated, researchers have proposed a bewildering array of practical applications, from faster computer chips, and flexible touch screens to super-efficient solar cells. It really is an amazing material, and little wonder everybody is so excited about it.

Significantly, many researchers believe that Graphene can play a key role in the development of electric vehicle batteries, giving them the storage capacity to rival the best Internal Combustion Engines.

One of the first discoveries was that if conventional battery electrodes are wrapped in a graphene sheet there is a significant improvement in the performance of the battery.

Graphene can be incorporated into batteries that are light, durable and suitable for high capacity energy storage, as well as shorten charging times. Graphene can improve such battery attributes as energy density and form in various ways. Li-ion batteries can be enhanced by introducing graphene to the battery's anode and capitalizing on the material's conductivity to achieve improved performance. It has also been discovered that creating hybrid materials can also be useful for achieving battery enhancement. A hybrid of Vanadium Oxide (VO^2) and graphene, for example, can be used on Li-ion cathodes and enable quick charge and discharge as well as large charge cycle durability. In this case, VO^2 offers high energy capacity but poor electrical conductivity, which can be solved by using graphene as a sort of a structural "backbone" on which to attach VO^2 - creating a hybrid material that has both heightened capacity and excellent conductivity, having greater capacity than conventional batteries.

In addition to revolutionizing the battery market, combined use of graphene batteries and super capacitors could yield amazing results, like the noted concept of improving an electric car's driving range and efficiency.

The use of Graphene in numerous applications continues to grow, and the science behind its amazing properties is now well understood. There is however, one main issue remaining. How do you mass produce the stuff?

We don't believe it is an obstacle which will remain for long.

A Graphene sheet

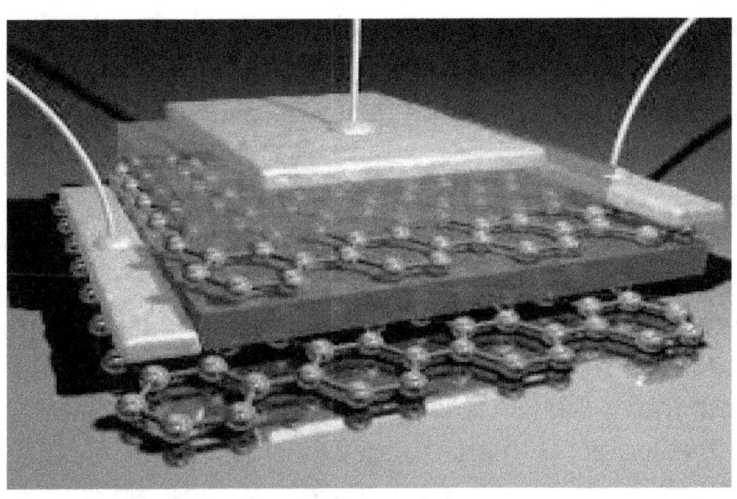

Lithium-ion batteries (**LIBs**) are currently used in the majority of electric vehicles, and it's likely that they will remain dominant for the immediate future. Several manufacturers, including Tesla, Nissan, Mercedes, General Motors and Ford are investing heavily in this technology. In LIBs, positively charged lithium ions travel between the anode and the cathode in the electrolyte. LIBs have a high cyclability – the number of times the battery can be recharged while still maintaining its efficiency. However, they do have a relatively low energy density i.e. the amount of energy that can be stored in a unit volume. LIBs earned a bad reputation in 2016 for overheating and catching fire, for example, Samsung mobile phones and Tesla cars. They were also implicated in the demise of one company, Fisker Electric Cars. Manufacturers have not only worked to make LIBs more stable, they have also developed safety mechanisms to prevent harm if a battery cell were to overheat. The LIBs initially available on the market used graphite or silicon anodes and a liquid electrolyte. A lithium anode has been the holy grail for a long time because it can store a lot of energy in a small space (*i.e. it has a higher energy density*) and is very lightweight. Unfortunately, lithium heats up and expands during charging, causing leaked lithium ions to build up on a battery's surface. These growths can short-circuit the battery and decrease its overall life. Worse still, it can cause a fire. Researchers at Stanford have made some headway on these problems by forming a protective layer on the lithium anode which enhances quick charge and discharge as, well as large charge cycle durability. In this case, VO^2 offers high energy capacity but poor electrical conductivity, which can be solved by using graphene as a sort of a structural "backbone" on which to attach VO^2 - creating a hybrid material that has both heightened capacity and excellent conductivity, having greater capacity than conventional batteries.

In addition to revolutionizing the battery market, combined use of graphene batteries and super capacitors could yield amazing results, like the noted concept of improving the electric car's driving range and efficiency.

Despite early day problems with over-heating, a major attribute of a Lithium-ion battery is its high recyclability and robust structure. Energy density is good and continuously being improved, and the prospect of large cost reductions.

To conclude, **the Lithium-ion** battery has become the most developed storage technology available today, and covers most cordless and mobile applications, including the power source for electric vehicles.

Dyson Technology.

Dyson Engineering, based in Malmesbury, Wiltshire UK is working on the development of a revolutionary new Lithium-ion battery for an electric vehicle.

The company has many years of know-how in volume production and is famed for its battery powered cyclonic vacuum cleaners, hand dryers, bladeless fans, heaters and hair dryers. It is well experienced in battery technology and now claims to have a battery suitable for a road vehicle. This is thanks to its $90m acquisition of battery company **Sakti3**, a start-up, launched from the University of Michigan by Professor **Ann Marie Sastry**. The **Sakti3 team has** published over 100 papers concerning super battery technology, and is a recognised international authority in this field of research.

Now Dyson, and the Sakti3 team **claim** to have developed a solid-state lithium-ion battery producing over 400Wh/kg energy density.- almost double the capacity of Tesla's Panasonic cells, reckoned to be the industry leader, and This equates to double the range. The Projected cost is $100 per kilowatt-hour, and that is the critical trigger point at which **electric** vehicles are claimed to equal fossil fuels in cost and range.

Today's lithium-ion batteries are typically packed out with gels or liquids that don't store any energy, but Professor Sastry claims they have discovered a 'solid' conductive material that is porous enough to let lithium-ions pass back and forth from anode to cathode, discharging and charging the battery. Her team then identified combinations of Lithium- ion materials and structures around lithium that would result in a high energy battery that could also be mass-produced affordably. " It's no use having the best energy density or greatest number of cycles if they are prohibitively expensive to manufacture" declares Professor Sastry.

In the prototype assembly of micro-thin layers that build up the battery, Sastry's team modified equipment used to make sheets of printed foil. In reality the same thin-film deposition process employed to make flat panel displays and photovoltaic solar cells that layer micro-thin films of cathode followed by the current collector, then the interlayer anode and so on, all within a vacuum. Once assembled the resulting cells are charged and ready for testing. Scaling up battery manufacturing from the laboratory to volume production is the innovation that will drive Dyson's breakthrough, say industry experts.

Lithium-sulfur batteries

Lithium-sulfur batteries (**Li/S**) typically have a Lithium anode and a sulfur-carbon cathode. They offer a higher theoretical energy density and a lower cost than LIBs. Initially, Their low re-cyclability, caused by expansion and harmful reactions with the electrolyte, was the major drawback, but has now been improved, and these batteries, combined with solar panels, powered the famous 3-day flight of the **Zephyr-6** unmanned aerial vehicle. Now NASA has invested in solid-state Lithium-sulfur batteries for use in space vehicles.

The new lithium–sulfur battery is notable for its high specific energy. The low atomic weight of lithium and moderate weight of sulfur means that Li/S batteries are relatively light. With an energy density of 400 Wh/kg,

Oxis Batteries

Based in Oxford UK, this company claims its batteries are nearly five times more powerful than its Li-ion counterparts. In addition they are lighter, safer and maintenance free. **Lithium Sulfur** batteries prevent fires and retain their functionality, even after accidents. ' Oxis' director, Dr Mark Crittenden, says that the company is aiming to capture 10% of the Lithium sulfur market in the next five years – a very ambitious target. To put this into context, data shows that Lithium Ion has grown from zero in 1990 to a forecast of $80 billion in 2020.

Aluminum-ion batteries

Aluminum-ion batteries are a new development of Lithium, but with an aluminum anode. They promise increased safety at a decreased cost over LIBs, but research is still in its infancy.

These batteries are a class of rechargeable battery in which aluminium ions provide the energy. They are conceptually similar to lithium-ion batteries, but possess an aluminium anode instead of a lithium anode.

While the theoretical voltage for aluminum-ion batteries is lower than lithium-ion batteries, 2.65 V and 4 V respectively, the theoretical energy density for aluminium-ion batteries is 1060 Wh/kg in comparison to lithium-ion's 406 Wh/kg

Disadvantages Aluminium-ion batteries have a relatively short shelf life, and combination of heat, rate-of charge, and cycling can dramatically decrease energy capacity. Also, when aluminium- ion batteries are fully discharged, it is extremely difficult to re-charge them. On the positive side, University research teams are working hard to find solutions to most if not all the above problems.

Lithium Air Batteries

These batteries are considered by some to be the ultimate battery due to their theoretical energy density, which is ten times that of an existing lithium-ion battery; and this makes them comparable to gasoline. Commercial lithium-air batteries could revolutionise the battery industry. They may enable electric cars to run on a battery that's a fifth of the cost and a fifth of the weight of batteries currently on the market. They could also give a range of 400 miles on a single charge.

However, there have been severe practical challenges impeding the development of lithium-air batteries, and previous attempts to develop a full-scale solution have resulted in low efficiencies and poor performance.

Now, after three years of concentrated effort, scientists at the **University of Cambridge in the UK,** have developed a working laboratory demonstrator of a lithium-oxygen battery with a high energy density, which is more than 90% efficient and can be recharged more than 2,000 times.

The results, published by Lead scientist, **Professor Clare Grey** in the prestigious peer-reviewed journal **Science**, claims the breakthrough solves the problems with current Lithium technology. Because lithium-air has such a big theoretical advantage over lithium-ion it could take over domination of the battery market. For example, Lithium-air's energy density is potentially 10 times greater, researchers around the world are working on lithium-air, so the paper published by the Cambridge team has triggered intense interest. The research paper shows that the Cambridge group has overcome some of the practical problems of the technology, particularly the chemical instability that led to a rapid fall-off in performance of lithium-air cells demonstrated previously.

The basic chemistry of lithium-air batteries is simple; the cell generates electricity by combining lithium with oxygen to form lithium peroxide. It is then recharged by applying a current to reverse the reaction. However, making these reactions take place reliably over many cycles is the challenge. The Cambridge scientists adjusted the chemistry to make it more controllable. For example, they converted lithium peroxide to lithium hydroxide (a compound that is easier to work with), they added lithium iodide to the system and they made a very porous "fluffy" electrode from graphene.

The system demonstrated in the Cambridge lab is 90 per cent efficient, claim the researchers, and it can be recharged 2,000 times. However, many years of work is likely to be required to turn it into a commercial battery for cars, and for the grid storage of intermittent quantities of wind and solar power.

The Cambridge research has been funded by the UK Engineering and Physical Sciences Research Council, the US Department of Energy and the EU, with support from Johnson Matthey, the UK advanced materials company."We have patented the technology and the intellectual property is owned by Cambridge Enterprise, the university's commercialisation arm," said Prof Grey. "We are working with a number of companies to take it forward."The research paper shows that the Cambridge group has overcome some of the practical problems of the technology, particularly the chemical instability that led to a rapid fall-off in performance of lithium-air cells demonstrated previously.

University of California *(UCLA)*

Researchers at the materials science and engineering department believe they are on to a solution using high-performance graphene-based electrochemical capacitors (EC's), that maintain excellent electrochemical properties under high mechanical stress. Richard B, Kaner professor of chemistry and materials science says, "our study demonstrates that graphene-based capacitors store as much charge as conventional batteries but can be charged and discharged 100 to 1000 times faster". This is a remarkable demonstration, blurring the distinction between super capacitors and batteries, which when realised in Commercial devices could lead the way to the next generation of battery storage systems in smart-phones, tablets, laptops and other personal portable electronics devices.

The technology could also have applications for batteries in Electric Vehicles.

University of Central Florida

Scientists from the University of Central Florida (UCF) have created a super capacitor battery that works like new even after being recharged 30,000 times. The research could yield high-capacity, ultra-fast-charging batteries that last over 20 times longer than a conventional lithium-ion cell. "You could charge your mobile phone in a few seconds and you wouldn't need to charge it again for over a week," says UCF postdoctoral associate Nitin Choudhary. Super capacitors can be charged quickly because they store electricity statically on the surface of a material, rather than using chemical reactions as in batteries. The new super capacitors require two-dimensional sheets of Graphene with large surface areas that can hold lots of electrons. However, Leading Research Scientist Eric Jung admits it is quite a challenge to integrate graphene with other materials used in these super capacitors. That is why the team wrapped 2D metal materials (TMDs) just a few atoms thick around highly-conductive 1D nano wires, letting electrons pass quickly from the core to the shell. The research is in its early days and not ready for commercialization, but it does looks promising. "For small electronic devices, our materials are surpassing the conventional ones in terms of energy density, power density worldwide, says professor Jung.

Tesla's Giant Battery Factory

Factory construction began in November 2014 The factory is being built in phases so that Tesla can begin manufacturing immediately inside the finished sections Other auto manufacturers, such as Toyota and Volkswagen, are also developing new solid state batteries to power their own electric cars.

SECTION 5

Charging and Storing

New and old methods to store potential power.

Storing Electricity

From the earliest days of electricity, the problem of storing excess power during the periods when supply exceeded demand, and then supplying it to the grid during those periods when demand exceeded supply, was and remains, a most important issue. Indeed, with the advent of intermittent sources of electrical power such as solar energy farms and enormous off-shore arrays of wind turbines Down **Time Storage** has become a critical issue.

Now Consider this:

The sun sets and the sky grows dark. The wind is calm.
Not even a breeze.
The result ? No Solar Electricity and no Wind generated electricity.

What do we do?

   Well at the moment it isn't much of a problem. Our electricity comes via the National Grid from both of the above sources plus Nuclear power, and the fossil fuel powered generators, i.e. Coal, Oil, and Gas. These are not reliant on Sun or wind, so the lights stay on. The result might be inconvenient, but not disastrous.
Now consider another scenario when, because of our international obligations to combat global climate change we have closed down all our fossil fuel burning power stations. This process is already in operation and, by the end of 2017 wind power alone was exceeding coal power. Without wind or solar we are reliant on nuclear power, and this alone is not enough to satisfy demand. The result is power interruptions.

Over the past 150 years, numerous ingenious methods have been devised to cover downtimes. these include Pumped hydro, Compressed air, Flywheel Lead acid, batteries, and Superconducting magnetic coils. All designed to keep the electricity flowing whilst the primary power source was down. Only two of these have managed to enter the mainstream of grid based regional power supply storage systems. They are: Pumped Hydro and Lithium-ion batteries.

In the following section we look at Ultra Large batteries and Pumped Hydro methods to keep the electric power on.

Batteries for Grid Scale Storage

A Storage Battery supplying 12 MW of power

Batteries come in many shapes and sizes, from miniature cells used to power hearing aids and wristwatches to battery banks the size of rooms that provide standby power for telephone exchanges and computer data centres. At the very top come batteries to provide backup power to electricity grids whose primary source of power is intermittent, such as in Solar arrays and Wind power. Both of these sources have periods of down time, in darkness or when the wind is calm, and it is in this application that Lithium-ion batteries are beginning to be employed. Indeed, they have become the most important grid storage technology that we have so far.

An important advantage of Li-ion batteries is their high electrical energy density, and the prospect of large cost reductions through mass production.
Lithium ion batteries generally have a very high efficiency, typically in the range of 95% power in to power out. Discharge times can range from seconds to many weeks, making them flexible and a universal storage technology. Standard cells with

5,000 full charge/discharge cycles can be obtained, and even higher cycles are being developed

There are Problems. Although Li-ion batteries have a share of over 50 % in the portable equipment markets, there remain serious challenges to employing them for use in electrical power distribution on a national grid. The main obstacle here is the high cost of the special batteries required; often amounting to millions of dollars per battery array. In addition, there is a critical issue of safety. As we have seen earlier the metal oxide electrodes used in a lithium ion battery are inherently thermally unstable, and can decompose at elevated temperatures. This causes the release of oxygen leading to the possibility of thermal runaway, and fire. To minimize this risk, lithium ion batteries are equipped with a built in monitoring unit to avoid over-charging and over discharging. In addition voltage balance circuitry is installed to monitor the voltage level of each individual cell, and prevent voltage deviations at elevated temperatures.

The technology is still developing, and although they are considered robust and reliable enough in the smaller applications, there is considerable potential for further advances in their employment at the top end.

An example of a live Grid Scale Storage System.

The Chinese company **BYD** operates a battery bank in Hong Kong with 40 MW/hour capacities and 20 MW maximum power. This large storage is used to cushion load peaks in energy demand. Likewise, the storage can contribute to the frequency stabilization in the network. The battery is made up of a total of almost 60,000 individual Lithium–ion phosphate cells, each with 230 amp hour capacity. The project was started in October 2013, and went live to the grid in June 2014. The actual installation of the storage elements took three months. The use of price differences between loading and unloading for grid services, avoided grid expansion for peak loads, and for grid services.

The heart of the electric vehicle is of course the battery, and by 2030 these will be produced on a mammoth scale. They will be cheap, reliable, safe, and fast charging. The standard vehicle battery will almost certainly have a better than 1,000 times recharge cycle life span, and under most driving conditions the power to provide a range of 500 km. Giving them the same operational characteristics as the old 'Internal Combustion Engine'. As mentioned earlier, we think cars will come with 2 batteries installed, both clipped into a cradle located in a separate compartment. The main active battery and 1 standby battery on trickle charge. These two batteries can be switched over from charging to active mode by a switch on the dashboard. Giving a total range of 1,000 Km, (or 600 miles). single battery. Indeed, an optional third battery could be located at home. It could then be left charging from home generated electricity.(*Solar, Biofuel Air-source heating or Ground source heating*). It will effectively be free fuel, with only 'Road Charges' and the Vehicle License, to worry about.

Charging Electric Car Batteries

The limited driving range of many electric vehicles means that the type of technology used to charge the battery, and the time it takes, is fundamentally important to users. All mass-produced electric cars today come with a battery charging plug and cable to connect to a standard 110-volt or 220 volt outlet. It is thus possible to charge a car battery from an ordinary household supply. A large reduction in the price of fuel. There are three basic ways to charge the car battery: - **(1)** plug-in charging, as above; **(2)** battery swapping or **(3)** cordless charging.

Plug-in charging is the most common, and the vast majority of Electric Vehicles in Europe, and the United States use this method. The car is plugged into a charging point using the cable and plug supplied by the manufacturer. It can be used wherever charging facilities are located:- at homes, and many thousands of 'Charging Points' springing up everywhere. However, this can be a rather slow method taking up to eight

hours for a full recharge. **Battery swapping** involves replacing a used battery with a fully charged one in a few minutes, at a garage, or fuel filling station, offering this service. It is still very rare or non existant throughout, Europe, but more common in North America. Bottled Calor Gas, or Propane Gas exchange services are however widely available in Europe, and can be expected

to extend to Car Batteries when electric vehicles become widespread.

A venture that sadly failed was described in the section on. **'Companies that did not Survive'. Better Place inc'**, was an Israeli company which launched a battery exchange service for purchasers of its electric cars. When a new battery was required, the car was driven into one of 37 automatic switching stations where the old battery was removed and replaced with a fully charged new one -- All within five minutes. It was a brilliant idea, but there were simply not enough customers in Israel to make the service viable, and the company failed.

Cordless charging.
A row of electric coils spaced just under the road surface can couple in sequence with an identical coil fixed underneath a vehicle passing overhead, and transfer an electric charge to the moving coil. If the coils all share exactly the same resonant frequence then significant charge can be transferred to the car battery. This method of charging works for moving and stationary vehicles.

We look further at this exciting development known as induction charging, later in this chapter

Recharging
Imagine never going to a fuel station again. All you have to do is pull into your garage or drive-way, reach over for a plug, and push it into the charging inlet. It's very convenient and takes all of 15 seconds. Charging takes place through the night on off-peak lower cost electricity. Wake up the next morning, and you have a car ready to go another 500 Kilometers (300 miles), and at a cost of 10% (*yes 10%)* of the price of regular fuel from the pump.

Home Chargers

There are plenty of Home Chargers available, and you can find them Online or from Amazon or Ebay.

Concerns about range are closely tied with issues related to how long it takes to refuel an electric car. The latest EVs can add about 60 to 80 miles of range in under an hour of charging from a standard mains voltage source of electricity. So whilst it is still not as easy to add a couple hundred miles of range in five minutes, it can be very cash efficient to plug in before retiring in the evening.

However, another factor to consider is public DC Quick Chargers, capable of adding 80 miles of range in around 20 minutes, are becoming increasingly available.

With the current rate of technology growth, we think that before 2030 cars will come with 2 batteries installed, both clipped into a cradle located in a separate compartment: the main active battery and a standby battery. These two batteries can be switched over from charging to active mode by a switch on the dashboard. Giving a projected range of 1,000 Km, or 600 miles. 'Range Anxiety', will finally have vanished for ever.

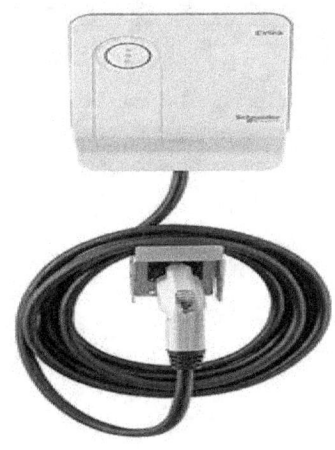

Solar Power Recharging. Options for fuelling your electric car with Free Solar power.

Home based
battery charging kit

Mains power rechargeing bay

A question that must be addressed before
making a long journey by Electric car is: --
"will I reach my destination before I run out of fuel, or will I have

to stop and wait for the battery to recharge"?

Even though some models can boast a 300 mile range before a recharge, it still then takes up to 8 hours to fully top up the battery.

Now wireless vehicle battery charging that could see electric cars get their batteries topped up while they are driven along the road: may be the answer, and closer than you think.

French Smartphone chip maker Qualcomm, Technologies has created a stretch of road in France that sends charge to electric vehicles travelling along it, even at high speeds. The company said it had "designed and built a wireless system capable of charging an electric vehicle dynamically at highway speeds". The introduction of this kind of wireless charging on highways, could revolutionise the entire electric-car market, eradicating 'range anxiety' and the requirement to stop for extended periods to recharge batteries. This amazing stretch of road incorporates a string of wireless charging pads that can replenish the batteries of electric cars as they move over it. The first live demonstration of Qualcomm's modified road has taken place in France, with some impressive results; clearly demonstrating that magnetic resonance coupling between a coil attached to the underside of the car, and identical coils with the exact same resonance frequency characteristics hidden just below the surface along the road The two coupled circuits allow transference of significant amounts of energy across the void putting charge into the car battery. All this whilst driving at speeds of up to 100kmh (62mph).

Most observers agree that this French breakthrough could make electric vehicles much more popular, and range anxiety finally eliminated. The ability to top up whilst moving will be a significant step foreward.

Experts at Qualcomm, Technologies have already suggested implementing vehicle charging in towns. Charging pads could be located into sections of road at traffic lights, car parks, and even in taxi ranks. Indeed, anywhere that traffic is stationary or parked.

The company believes that the introduction of in-town wireless charging facilities could further encourage the use of electric vehicles in urban areas. The environment would benefit. It would becomes quieter, with clean air; making it a more pleasant place to live.

It is early days yet, and so far only a few pilot locations exist. However, there are examples of inductive charging for buses at bus stations in Belgium, Germany, the Netherlands and the United Kingdom, as well as some pilot testing for users of electric cars in Sweden.

Practical Deployment

A single vehicle-width recharging lane is constructed alongside the main highway. Vehicles can enter one way and depart at its far end.

Just underneath the surface of the recharging lane is a string of wireless charging pads (*like those shown above*) that can replenish the battery of an electric car that is fitted with an identical electromagnetic coil ,and is driven along the lane.

The system also works by placing wireless charging pads

underneath the tarmac at traffic lights, car parks and taxi ranks in order to give charge to an electric car battery when the vehicle hovers stationary or is parked over it.

There is little doubt that introducing wireless charging on this scale, on highways, and in the town could revolutionise the electric-car market, making it more convenient, and potentially cheaper than using fossil fuels, whilst removing the impracticality of having to stop for extended periods to recharge batteries: - two of the biggest hurdles for electric vehicle adoption today.

Time to get technical

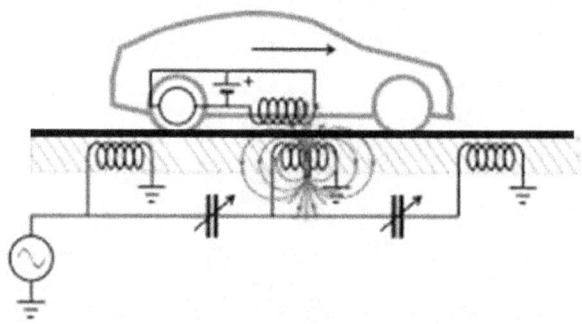

The scientific principle central to this subject was named **'Magnetic induction'** by its discoverer Michael Faraday at the beginning of the 19th century. Inductive coupling, uses the alternating magnetic field created by an alternating current (ac) surrounding a primary coil to induce an ac current in a closely adjoining secondary coil with no physical contact between the two. This works fine in Electric transformers, and is capable of transferring charge to the tiny batteries of items such as Smart-phones, although it is too weak to transfer significant charge to much larger batteries such as those required for electric vehicles. However, by employing a second phenomena, called magnetic resonance, inductive coupling can be much higher allowing sufficient charge to be transferred from one coil to the other; thus enabling a large battery to receive significant charge.

Stopped up. The key here is to configure the two coils so that both have exactly the same 'natural' resonant frequency. This can be a difficult operation when one coil is lying just below the tarmac whilst the other is fixed underneath the vehicle, and moving across the gap between them.

Despite the delicate fine tuning required, the system can

be made to work well. In fact this configuration has kept 30 electric buses running in Genoa and Turin, Italy for more than a decade.

 Coils buried in the road restore 10 to 15 percent of the bus battery's charge during each stop for passengers.

These prototype wireless charging systems map real progress towards the ultimate aim of an in-motion a recharging option for EV's along all main highways.

Charging
Pad

SECTION 6

How we Generate Electricity

It's a fact:

The sun provides more than enough energy in just one hour to supply our planet's whole energy needs for a year.

It is self evident that a large electricity power station requires a large amount of mechanical power to turn the giant Turbines, often weighing several tons. The turbine blades are made to rotate by the force of super-heated steam, and this in turn rotates an **Armature** upon which coils of **copper wire** are encircled. It is, only by cutting through magnetic lines of force that invisibly small electrons in the outermost orbits of the wire's **copper** atoms will be torn away from their parent and flow like a river towards the positive terminal of an electric circuit; even when that circuit is hundreds of miles long. The power to generate the steam can come from a variety of sources, including **Coal, Oil, Gas,** or **Biomass**, which is burnt in a furnace that boils water to produce super-heated steam. In a nuclear power station there is no furnace, and the heat to drive the turbine comes from a controlled nuclear chain reaction. Other common sources of mechanical power required to turn the turbine, come from Waterfalls, the Wind, the Waves, or Tidal flows. These work by directly applying their force against the turbine blades. Hot water is not required, and there are no greenhouse gases produced, as in the case of burning fossil fuels.

Armature

Solar Electricity

A common Domestic rooftop solar array in Britain

Nearly all electricity is generated by rotating copper wires through a magnetic field. The magnetism attracts loosely held electrons, which are orbiting far out from the nucleus of their copper atom, and pulled out of their orbit, into a flowing stream of electrons, speeding towards a positive charge on a distant terminal. This is an electric current, and this is how electricity is made. Almost all electricity relies on this rotating coil of copper

wires to get an electric current to flow. In a power station this rotation is produced by super-heated steam, wind energy or Water energy driving round the blades of a giant turbine.

However, There is one kind of electricity that is different from all the rest. One that doesn't need a turbine, and the only one that Michael Faraday, who invented the electric generator (*and motor*), 200 years ago would not have recognised. That exception is Solar Electricity.

Please note: a battery is a device that stores sub-atomic particles, and releases them as electric current on demand. It doesn't make electricity.

How it works

Some materials exhibit a property known as the photoelectric effect that causes them to absorb photons of light and release electrons. When these free electrons are attracted towards a positive charge in a circuit, an electric current results. There are no moving parts, and no need for turbines driven by superheated steam or pressure from the wind or water. There are also no polluting greenhouse gasses.

Solar power is arguably the cleanest method of producing electricity presently available.

Photovoltaics (PV) is the science of direct conversion of light into electricity at the atomic level. The photoelectric effect was first noted by a French physicist, **Edmund Bequerel**, in 1839, who found that certain materials would produce small amounts of electric current when exposed to light. In 1905, **Albert Einstein** described the nature of light and the photoelectric effect

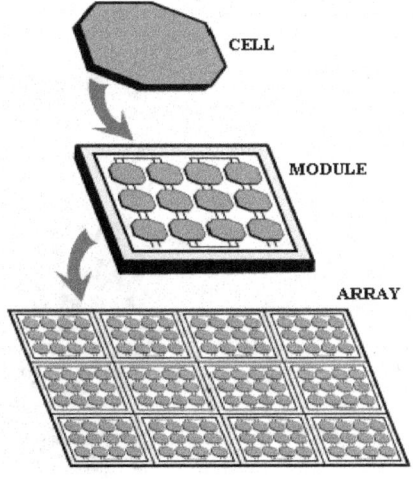

on which photovoltaic technology is based. For this he won his first Nobel Prize in physics.

Photo voltaics were initially used as a source of electricity for small and medium-sized applications, such as calculators, watches and toys, powered by a single solar cell, but as the cost of solar cells fell, groups of them were electrically connected to each other and mounted in **a** frame called a photovoltaic module. These modules are designed to supply **direct current electricity** at a certain voltage, such as a common 12 volts system. Multiple modules can be wired together to form an **Array**. In general, the larger an array, the more electricity it will produce. Large array systems can total thousands of modules, and produce Mega watts of power. After being converted from DC to the correct AC voltage, the electricity is transferred to the national grid without the need for turbines or any moving parts. The number of grid-connected solar PV systems has grown into the millions and utility-scale solar power stations with hundreds of megawatts are being built. Solar PV is rapidly becoming an inexpensive, low-carbon technology to harness renewable energy from the Sun. A typical photovoltaic system employs banks of solar panels, each comprising a number of solar cells, which generate electrical power.

A Solar Farm in East Anglia Great Britain

On balance Solar power is considered to be a positive thing; environmentally free, and Infinite. However, there are downsides : A Solar Energy Power Station (or Solar Farm) feeding Mega Watts of electricity into the national grid occupies a large area of land directly exposed to sunlight. The fierce heat of the tropics will downgrade the efficiency of the solar panels as will the dry dusty environment of a desert. Solar cells do not produce electricity during the periods of darkness. In cloudy or foggy weather the array will produce much less electricity. However, despite these limitations, the International Energy Agency has projected that, solar photo voltaics will become the world's largest source of environmentally, non polluting electricity for a lengthy period in mid 21st century. It also says that, during this period, the majority of solar installations will be in China and India.

Space Based Solar Power (SBSP)

Soon after solar power generation was developed in the late 1960's it became apparent that if a large array of Photovoltaic (PV) cells could be placed some 35,000 kilometers high above the earth's atmosphere in a geo-stationary orbit they could generate much more electricity per cell than on earth. Furthermore, this would be for almost 24 hours a day. Indeed two panels spaced slightly apart could avoid all of the earth's shadow and operate around the clock in perpetual sunlight. The electrical energy generated in each solar cell of the giant array is combined into one large electric current (DC), in the same way as is done in earth based Solar Farms. The combined electric current is then fed into a high power micro-wave transmitter, of a similar kind to those used in radar systems and micro wave ovens. The high power beam of micro waves is then transmitted down to earth to be picked up by a large cluster of wireless aerials. If the micro-wave beam is spread out in a cone across some 20 square kilometers it would not be harmful, or even noticeable by humans or wildlife, including birds. The tiny electric current generated in each aerial can then be combined into one large electric current which is fed to the power lines of the National Grid.

In 1973 a Mr Peter Glaser was granted U.S. patent (**3781647**) for his method of transmitting microwave power from a satellite to Earth's surface. A very large solar panel array of up to one square kilometre would first of all be used to collect the solar energy from the sun, to be fed into Mr Glaser's microwave power transmitter. On earth below, an array of thousands of tiny aerials called rectenna's covering 4.5 x 4.5 kilometres collect the microwave power and convert it into the correct electricity to be fed to the national grid.

The invention of the rectenna was patented in 1969 by US electrical engineer William Brown So attractive was this proposal that Nasa carried out a detailed examination of the concept, which concluded that **Space Based Solar Power (SBSP)** was a viable technology with applications not just on earth, but could also be used for supplying bases on the moon with electrical power. In its report, Nasa also identified specific advantages and disadvantages of SBSP, and these included the significant increase, from a terrestrial 15% to 60%,efficiency in solar energy collection due to the lack of atmospheric absorption and cloud reflection. Also the much longer collection period in space. Part of present day solar energy (55–60%) is lost on its way through the atmosphere by the effects of reflection and absorption. Space-based solar power systems convert sunlight to microwaves outside the atmosphere, avoiding these losses, and the downtime caused by the Earth's rotation. Besides the cost of implementing such a system, SBSP also introduces several new hurdles, primarily the problem of transmitting energy to Earth's surface.

The concept of SBSP is being actively pursued by Japan and China, and as long ago as 2008, Japan passed its Basic Space Law which established Space located Solar Power as a national goal with a road map to commercial realisation. In 2015 the China Academy for Space Technology (CAST) announced their own plans to Develop Solar Power and showcased their road map to a 1 GW Watt commercial system by 2050, and in 2015 unveiled a video describing of the concept.

In 2015 a proposal for the United States to lead in Space Solar Power received high level attention by the US Government.

Advantages of Satelite Based Sola Power

Microwave power transmission of tens of kilowatts was first proven to be feasible in tests between Reunion Island and
Maui on the island of Hawaii (92 miles away.

Space based solar power systems appear to possess many significant environmental advantages when compared to some of the disadvantages of terrestrial solar energy farms. These disadvantages include: daytime only operation, lower sunlight to electric current efficiency conversion, and higher PV panel surface contamination. Many researchers around the world now believe that Space based solar power generation may well emerge as a serious candidate among the options for meeting the energy demands of the 21st century. The May 2014 IEEE Spectrum magazine carried a lengthy article titled "It is Always Sunny in Space" by Japanese Space Scientist, Dr. Susumu Sasaki. The article stated, "It has been the subject of many previous studies and the stuff of sci-fi for decades, but space-based solar power could at last soon be reality—and that is within 25 years". A step further towards this was On 12 March 2015 when scientists at the Japan Aerospace Exploration Agency (JAXA), announced they had wirelessly beamed 1.8

kilowatts 50 meters to a small receiver by converting electricity to microwaves and then back to electricity.

As has previously been stated in this section, there are many serious and very practical advantages of **SBSP:** Not least among them are:- In space collecting surfaces could receive much more intense sunlight, owing to the lack of obstructions such as atmospheric gasses, clouds, dust and other weather events. Consequently, the intensity in orbit is approximately 144% of the maximum attainable intensity on Earth's surface. A satellite solar array could be illuminated over 99% of the time, and be in Earth's shadow a maximum of only 72 minutes per night at the spring and fall equinoxes at local midnight. Orbiting satellites can be exposed to a consistently high degree of solar radiation, generally for 24 hours per day, whereas the average earth surface solar panels currently collect power for an average of only 29% of the day. It is quite conceivable that in particular circumstances Power could be quickly redirected to areas that need it most. A collecting satellite could possibly direct power to different surface locations based on geographical peak load power needs. Typical contracts would be for base-load, continuous power, since peaking power is the main advantage of locating a space power station in geostationary orbit. This is because the antenna geometry stays constant, and so keeping the antennas lined up is simpler. Another advantage is that nearly continuous power transmission is immediately available as soon as the first space power station is placed in orbit; other space-based power stations have much longer start-up times before they are producing nearly continuous power. It is entirely possible that mining and manufacturing facilities on the Moon could be serviced by **SBSP.**

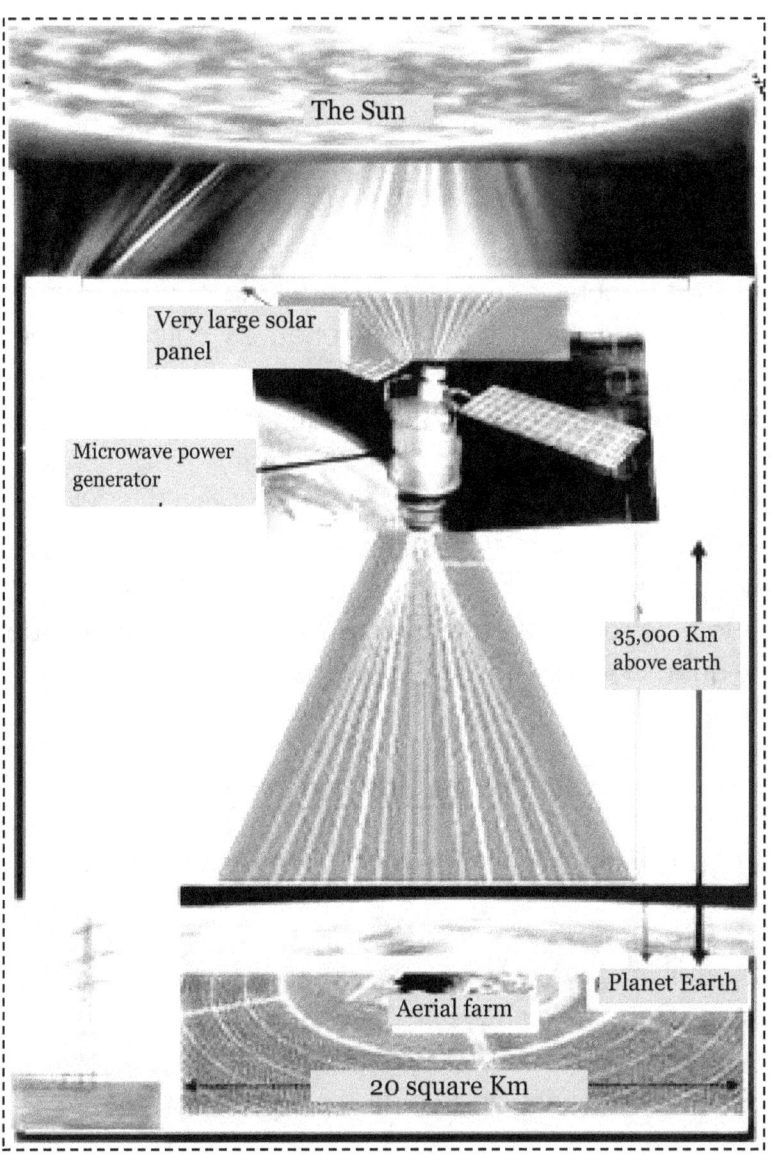

Disadvantages of Satelite Based Sola Power

The SBSP concept also has a number of problems: The space environment is hostile; panels suffer about 8 times the degradation they would on Earth (except at orbits that are protected by the magnetosphere. Space debris is a major hazard to large objects in space, and all large structures such as SBSP systems have been mentioned as potential sources of orbital debris. The broadcast frequency of the microwave downlink (if used) would require isolating the SBSP systems away from other satellites. Geo-Stationary- Orbit space is already well used. The high cost and size, possibly 20km², of the ground receiving station could also be an issue. The energy losses during several phases of conversion from "photon to electron to photon back to electron, has not yet been properly quantified.

Solar cell efficiency

The factors affecting Solar Energy conversion efficiency

Solar cell **efficiency** refers to the portion of energy in the form of sunlight that can be converted via photovoltaics into electricity. The **efficiency** of the **solar** cells used in a photovoltaic system, in combination with latitude and climate, determines the annual energy output of the system.

For example, a solar panel with 20% efficiency and an area of 1m2 will produce 200 W under 'Standard Test' Conditions, although it can produce more when the sun is high in the sky, and less in cloudy conditions or when the sun is low in the sky. In central Colorado, the amount of solar radiation reaching a solar panel can be expected to produce 440 kWh of energy per year. However, in Michigan, less solar radiation reaches the panel and annual energy yield will drop to 280 kWh for the same panel. At more northerly European latitudes, yields are significantly lower. For example 175 kWh annual energy yield in southern England.

As of December 2014, the world record for solar cell efficiency at 46% was achieved by using a multi-junction concentrator. Solar cells, developed by a collaboration between f Soitec, CEA-Leti, in France and Fraunhofer ISE, in Germany. However, there is a way to boost solar power, by increasing the light intensity. Typically photo-generated carriers are increased, by up to 15%. using a so-called concentrator. These have become cost-competitive as a result of the development of high efficiency Gallium-Arsenide cells. (GaAs) The increase in intensity is typically accomplished by using concentrating optics. A typical concentrator system may use a light intensity 6-400 times the sun, and increase the efficiency of a GaAs cell from 31% to 35%.

<u>Technical methods of improving efficiency</u> Promoting light scattering in the visible spectrum_By lining the light receiving surface of the cell with nano sized metallic studs, the efficiency of the cell can be substantially increased, as the light reflects off these studs at an oblique angle to the cell, increasing the length of the path the light takes through the cell, thereby increasing the number of photons absorbed by the cell, and so also the amount of current generated. The main materials used for the nano studs are silver, gold, and aluminium, to name a few. However, gold and silver are not very efficient, as they absorb much of the light in the visible spectrum, Aluminium, on the other hand, absorbs only ultraviolet radiation, and reflects both visible and infra-red light, so energy loss is minimized on that front. Aluminium is therefore capable of increasing the efficiency of the cell by up to 22%. <u>Radiative cooling</u> An increase in solar cell temperature of around 1 °C leads to a

decrease in efficiency of about 0.45%. To prevent decreased efficiency due to heating, a transparent silica crystal layer can be applied to a solar panel. The silica layer acts as a thermal black body which emits heat as infrared radiation into space cooling the cell. Aluminium, on the other hand, absorbs only ultraviolet radiation, and reflects both visible and infra-red light, so energy loss is minimized on that front. Aluminium is therefore capable of increasing the efficiency of the cell by up to 22%.

POWER from the WIND

Wind turbines operate on a simple principle. The energy in the wind turns two or three propeller-like blades around a rotor. The rotor is connected to the main shaft, which spins a generator to create electricity

Wind turbines come in a wide range of sizes with small 10 metre home garden models for battery charging or home power support to super size 350 metre off-shore monsters supplying power to the  national grid. At a national level, super large turbines, are becoming an increasingly important source of intermittent renewable energy, and are used by many countries as part of a strategy to reduce their reliance on fossil fuels. Wind was shown to have the "lowest relative greenhouse gas emissions, the least water consumption demands and... the most favourable social impacts compared to photovoltaic, hydro, geothermal, coal, gas and oil. The United Kingdom is one of the best locations for wind power in the world, and is considered to be the best location in Europe. Wind power contributed 11% of UK electricity generation in 2015, and 17% in December 2015. Allowing for the costs of pollution, particularly the carbon emissions of other forms of production, onshore wind power is the cheapest form of energy in the United Kingdom. In 2016, the UK generated more electricity from wind power than from coal. Wind power delivers a growing percentage of the energy of the United Kingdom and by the end of January 2018, there were 8,554 wind turbines with a total installed capacity of over 17.8 gigawatts: 12,082 megawatts of onshore capacity and 5,788 megawatts of offshore capacity. This placed the UK world's sixth largest producer of wind power (after China, USA, Germany, India and. Spain. Overall, wind power raises costs of electricity by a modest amount. In 2015, it was estimated that the use of wind power in the UK had added £18 to the average yearly electricity bill.

Home Generated Wind Power

Small low cost wind turbines up to 10 metres high, like the one illustrated here, are now available to work in tandem with a solar panel for charging continuity during the hours of darkness. The one illustrated is mounted on a roof, and costs around US$ 450.

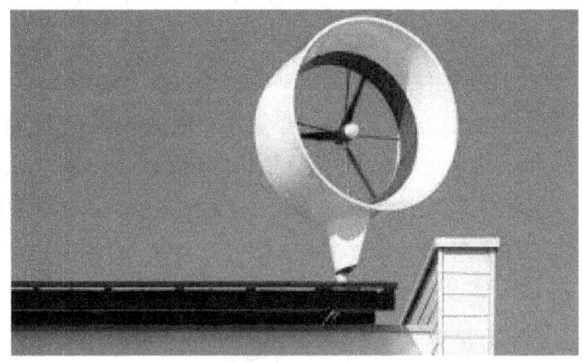

A typical off-shore wind Farm 40 Km off the UK coast

Hydro Electricity

In mountainous regions of the world where a plentiful supply of rainwater feeds fast flowing rivers, dams can be employed to generate electricity. This is called Hydro Electric power, and is the world's leading renewable energy resource and the oldest method of harnessing clean power - the first waterwheels were used over 2,000 years ago. Hydro currently produces around 17% of the world's electricity and 90% of the world's renewable power Hydro power does not produce any green house gas, and for regions of the world with the right terrain is a safe, reliable and very low cost method of producing electricity. Norway for example produces 99% of its electricity from hydro power plants. In 2015, hydropower represented 2.4 percent of the total energy consumed in the United States, whilst in the

United Kingdom this figure was just 1.5%.

Tidal Power

Tidal power or tidal energy is a form of hydro-power that converts the energy obtained from tides into useful forms of power, mainly electricity. Although not yet widely used, tidal energy has huge potential for future electricity generation. Tides are more predictable than the wind and the sun. Among sources of renewable energy, tidal energy has traditionally suffered from relatively high cost and limited availability of sites with sufficiently high tidal ranges. However there are a number of locations around the world with huge tidal ranges, making them eminently suitable for Tidal Power Stations. These include: The world's largest tidal range of 16.3 metres (53.5 feet) that occurs in the **Bay of Fundy,Canada**.
The Severn Estuary in the United Kingdom is number 2, and experiences tidal ranges up to 15 metres.

Other locations for extreme tides suitable for large power generation include:-
Cook Inlet, Alaska USA.
Rio Gallegos Argentina.
Magellan Strait, Chile
Granville   France.

Recent advances in turbine technology indicate that the total available tidal power may be much higher than previously assumed, and that costs could be substantially reduced. Historically, tide mills have been used both in Europe and on the Atlantic coast of North America. The incoming water was contained in large storage ponds, and as the tide went out, it turned waterwheels that used the mechanical power it produced to mill grain. The earliest occurrences of using falling water power to create electricity was introduced in the U.S. and Europe in the late 19th century.
The tides around Britain's coasts sweep huge volumes of water back and forth at substantial speeds

The energy contained in the tidal races off the west of the UK is as great as anywhere in the world. Because water is a thousand or so times heavier than air, the maximum speeds of perhaps 6 metres a second are capable of generating far more electricity per square metre of turbine area than a wind turbine. The Pentland Firth, the narrow run of water between the north-east tip of Scotland and the Orkney islands, is possibly the best place in the world to turn racing tides into electricity. The challenges are immense: massive steel structures need to be made that survive huge stresses, day after day. However, the rewards for tidal stream developers are commensurate. Unlike other renewable technologies, tidal power is utterly predictable for the entire life of a turbine. Once installed, the running cost of tidal stream technology is modest, and The UK could probably provide a quarter of its electricity from the tides.

Rance 240mW Tidal Power Plant, France -

The 240mW La Rance tidal power plant on the estuary of the river Rance in Brittany, France, has been operational since 1966 making it the world's oldest and second biggest tidal power station. The plant site features an average tidal range of 8.2m, the highest in France.  Electricity is fed into the national grid and supplies power for 130,000 households

Sihwa Lake Tidal Power Station, South Korea

With an output capacity of 254mW, the Sihwa Lake tidal power station in South Korea, was opened in August 2011, and is the world's biggest. It utilises a 12km long seawall, (originally, constructed for flood protection), to trap tidal seawater. Power is generated on tidal inflows into the 30km^2 basin with the help of ten 25.4mW submerged bulb turbines, and 8 sluice gates controlling water outflow from the barrage

Dynamic tidal power A new tidal energy design option is to construct circular retaining walls embedded with turbines that can capture the potential energy of tides. The created reservoirs are similar to those of tidal barrages, except that the location is artificial. The lagoons can also be in double (*or triple*) format without pumping or including pumping that will even out the power output. The pumping power could be provided by excess-to-grid demand or renewable energy from, (*for example*) wind turbines or solar photovoltaic arrays. Excess renewable energy rather than being curtailed could be stored in the arrangement for use at a later time. Geographically dispersed tidal lagoons with a time delay between peak periods would also flatten out production. The proposed Tidal Lagoon in Swansea Bay Wales, UK, is one such example, and would be the first tidal power station of this type.

The tidal lagoon is a 'U' shaped breakwater, built out from the coast which has a bank of hydro turbines in it. Water fills up and empties the man-made lagoon as the tides rise and fall. Electricity is generated on both the incoming an outgoing tides, four times a day, every day. Due to very large tides on the West Coast of Britain, turbine gates need to be shut for only three hours, to attain a 4m height difference in water between the inside and the outside of the lagoon. Power is then generated as the water rushes through 60m long draft tubes, rotating the 7.2m diameter hydro turbines. The project was awarded a Development Consent Order in 2015 and is now under construction. It will comprise 16 hydro turbines, and a 9.5

km breakwater wall, generating electricity for 155,000 homes for the next 120 years.

<u>Swansea Bay Tidal Lagoon</u>

The 240mW Swansea Bay Tidal Lagoon project, is the world's first of its kind. The plant will be located at a site with an average tidal range of 8.5m and will involve the construction of a 9.5 km-long breakwater facility that creates a lagoon cordoning off 11km^2 of sea. The plant will use reversible bulb turbines to generate power as water passes in and out of the lagoon with the rise and fall of tides. Having an estimated annual power generation capacity of 400 GWh. It will power over 120,000 homes for 120 years.

Wave energy

Wave energy, is essentially power drawn from waves. When wind blows across the sea surface, it transfers energy to the waves. They can then become a driving force for a turbine and generate electricity.

The energy output is measured by wave speed, wave height, wavelength and water density. The more strong the waves, the more capable it is to produce power. The captured energy can then be used for electricity generation, powering plants or pumping of water. It is not easy to harness power from the waves or build robust enough generator plants to withstand the stormy ocean. For this reason there are very few wave generator plants around the world.

SECTION 7

Fossil Fuels

Primeval Forests

There are three major forms of fossil fuel: Coal, Oil and natural Gas. All three were formed some 300 million years ago in the so called Carboniferous period. Carbon, being the basic element in all three of these fossil fuels. During this period much of the land was covered with swamps, and densely populated with huge tree-like, ferns and other large plants. It is from this period that the main deposits of coal, oil and gas originate. As the trees and plants died, they sank to the bottom of the swamps or the oceans. Here they were covered with layer upon layer of peat or sand slowly building up, and pressing everything further down into the earth. Coal, oil and gas, are the fossil fuels responsible for the majority of the world's electricity and transport demands.

SMOG ! What is it?

Fossil fuel power stations are those that generate electricity by burning coal, oil, or gas, to produce super-heated steam. which turns the blades of a large turbine.Unfortunately, a large amount of carbon dioxide, (CO_2)Green House gas, is released into the atmosphere by burning thise ancient carboniferous materials Victorian London was notorious for its thick smogs, or "Pea-Soupers", they were called, from the phrase "as thick as pea soup." There are two recognized types of smog, and they have been around since the Middle Ages. There is sulfurous smog and photochemical smog.Sulfurous Smog puts a high concentration of Sulfur oxides in the air, whilst Photochemical Smog has nitrogen oxides and hydrocarbon vapors nowadays mainly emitted by automobiles As far back as 1306, concerns over air pollution were sufficient for King Edward I, to try and ban coal fires in London. The "Ballad of Gresham College" describes how the smoke "does our lungs and spirits choke". Severe episodes of Smog continued through the 19th and into the 20th century, mainly in the winter. The Pea-Souper Great Smog of 1952 caused the deaths of thousands of people, especially those with existing bronchial illness. In the early 1960's the 'Clean Air Act' started to be legally enforced. Smokeless zones in the capital were those areas where no soft coal was allowed to be burned in homes and businesses. Coke and Anthrocite, which produce no smoke, were permitted. Smog has now disappeared from London, although many other great cities are still effected by it. These include Mexico City, New Delhi, and Beijing.

> The following is a personal account of the last Pea-Souper fog that descended on London Thames Valley On Monday, December 3rd 1962.

At about mid morning, a dense 'Chemical Smog settled over London and the whole Thames Valley area, enveloping us all in a horrible, choking, foul tasting fog. It did not clear until Friday, five days later. The result was that visibility was down to a metre or the daytime. Walking out of doors became a matter of shuffling one's feet to feel for road curbs, etc. This was made even worse at night because street lamps were mostly fitted with non light penetrating bulbs and you couldn't even see your feet. 'Smog masks' were worn by those who were fortunate to be able to get one from the chemists, who were soon sold out, but they were not very effective, and did nothing to alleviate the stinging eyes. I rather unwisely tried to drive the three miles home from my company Decca Radar Ltd, to Kingston, but it was so dangerous I quickly had to abandon my car down a side street, and walk back. This is where the car stayed all week. I made the journey by bus. Even bicycles were too dangerous to ride, but the buses had powerful, low placed fog penetrating headlights, and were the only transport that worked. Five days later, when the fog had lifted, I had to search the side streets to find my car. During the five days it lasted, some 4,000 people with lung problems are reported to have died. Afterwards the **'Clean Air Act'** was rushed through parliament, and I don't believe there has been a pea soup fog in the Thames Valley since.

COAL

This is a calcified combustible carbonate material of black or brown sedimentary rock which is what remains of prehistoric forests, some 300 million of years old, that have been subjected to heat at enormous pressure deep within the earth. Coal plays a vital role in electricity generation worldwide, and in 2014 coal-fired power stations fuelled 41% of global electricity. However, coal is a major producer of greenhouse, gas as mentioned previously, and according to a consensus of opinion from environmental scientists, is the major man-made contributor to global climate change. In addition to its Greenhouse gas emissions, burning coal produces some other harmful by products. Not least is the generation of large quantities of mercury, sulfur dioxide, carbon monoxide, mercury, selenium, and arsenic. These substances not only cause acid rain, but are also toxic to humans and animals.

Burning coal to provide heat began in ancient times, and has also been used for cooking, smelting and pottery making. Coal still remains plentiful, and is readily available throughout the world. Britain still has huge deposits of coal beneath its hills with an estimated 1,500 years of deposits still readily accessible, and coal is likely to remain competitive for more years yet. Finally, coal is considered a safe fuel, compared for example to nuclear power. The failure of a coal fired power plant is certainly not likely to cause catastrophic events such as in a nuclear meltdown.

However, the climate is changing for old King Cole. In 2017, figures produced showed the number of coal fired power stations under progress worldwide had almost halved, as international concern for environmental issues turned against it. 570 gigawatts of coal plant capacity was in the pre-construction pipeline as of January 2017 down from 1,090 gigawatts a year earlier the study by the environmental groups **Greenpeace** and the **Sierra Club** reported. There has been a dramatic clampdown on new coal plant projects by the Chinese central authorities and financial cutbacks for coal plant development in India. The report concluded, "We have perhaps seen a turning point, curbing the burning of coal for electricity generation". Such a development is essential if the world is to achieve the emissions reduction targets enshrined in the 2015 Paris climate change agreement. This seems quite likely to have put the climate goals within reach, preventing irreversible climate change.

Drax Power Station

Oil

Another fossil fuel is Oil, which was formed more than 300 million years ago from tiny sea creatures (*the size of a pin head*), called Diatoms. Like plants they convert sunlight directly into stored energy. The sea is still home to enormous numbers of these tiny organisms which are known as phytoplankton, and are still at work today.

As the diatoms died they fell to the sea floor and were buried under the sediment to become carbonized under great pressure and heat, eventually being turned into oil. As sea levels rose and fell, and the earth moved, up and down, folded, pockets where oil and natural gas collected were formed, sometimes close to the surface, and sometimes kilometres deep below it.

Oil has been used for more than 5,000 years. The ancient Assyrians and Babylonians used crude **oil** and asphalt ("pitch") which they collected from large seeps at modern-day **Heet,** an ancient town in Iraq on the banks on the Euphrates River. A seep is a place where the oil leaks up from below ground. The ancient Egyptians, used liquid oil as a medicine for wounds, and oil was used in lamps for light.

The Dead Sea, in Israel, used to be called Lake Asphaltites. Where gooey lumps of the stuff was washed up on the shore line.

An Oil Seep in California.

Oil is made into many different products – fertilizers for farms, the clothes you wear, the toothbrush you use, the plastic bottle that holds your milk, the plastic pen that you write with. They all came from oil. There are thousands of other products that come from oil. Almost all plastic comes originally from oil. Other products include gasoline, diesel fuel, aviation or jet fuel, home heating oil, oil for ships and oil to burn in power plants to make electricity. It's quite amazing the number of things a barrel of crude oil can make.

A commercial oil field uses hundreds of oil pumps working around the clock. 24/7. Crude oil comes out of the ground as a dark, sticky substance, looking

6 Gallons of fuel looks more like Tar than a liquid. Before it can be used for just about any modern application it has to be changed, and this is done in an Oil refinery, with a process Cracking. The first step in refining crude oil is to heat it until it boils, when it can then be separated into different liquids and gases by a distillation process. These liquids are then transformed into the wonderful array of products we mentioned earlier.,

A single 42 Gallon Barrel of crude oil contains :

12 Gallons of Diesel,

19 Gallons of Gasolene

NOTE: If you prefer it you can skip this section
until Nuclear devices are explained.

An Introduction to atomic Physics

SECTION 8

What you should know about Electical theory

Atoms.

Scientists believe that all the matter in the universe is made up from tiny particles we call atoms. Atoms are the building blocks for making everything. Atoms are like tiny Solar Systems, and have at their centre a huge **Nucleus** which itself is made up of particles called **Protons** and **Neutrons**.

Electrons

These are subatomic particles, each some 1,000 times smaller than a Proton or a Neutron. They orbit the Nucleus of an atom (*think of the planets orbiting the sun*). Electrons in the far flung outer orbits of metal atoms such as lead, copper, silver and gold are not as tightly bound by the electro-static force to the massive atomic nucleus as those close to the centre (think gravity). These outer electrons can easily be detached by applying an external positive electric charge. Since electrons are deemed to be negative and therefore attracted towards a positive voltage charge, (*remember "like charges repell unlike charges attract*) those now free will flow like the water in a stream, towards a positively charged terminal.

Atoms can join together to form **Elements**, which themselves can group together to form You, and Me, and all the other objects which make up our world.

There are118 basic elements that have been identified, of which the first 94 occur naturally on Earth. Some like gold, silver, copper and carbon, have been known for thousands of years, others have only recently been discovered or fabricated by scientists.

Orbitting around the Nucleus are particles called Elect

An Atom

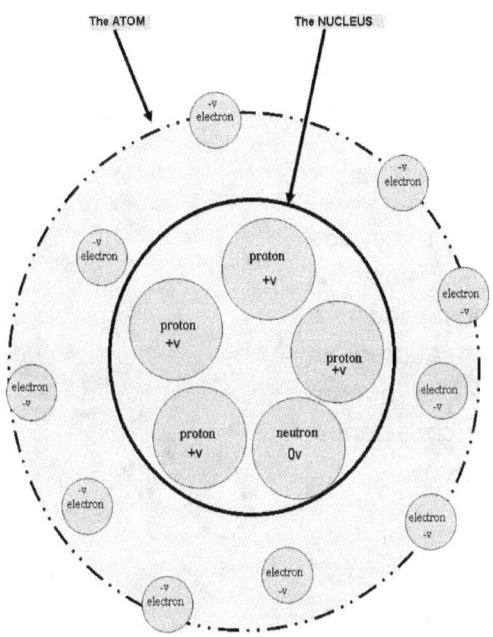

Copper Atom

When a terminal is connected in a circuit.
Electrons begin flowing, and this is an **Electric Current**.

Outer electron

An electric current is made up of many free electrons flowing like water down a pipe. This is measured in 'Amperes'; named in honour of the great French scientist **André-Marie Ampère**.

An electric current of **One amp** equals 6.241 x 10^{18} electrons flowing past a point in the wire in one second. That is **624,100**, trillion electrons per second.

Copper atom

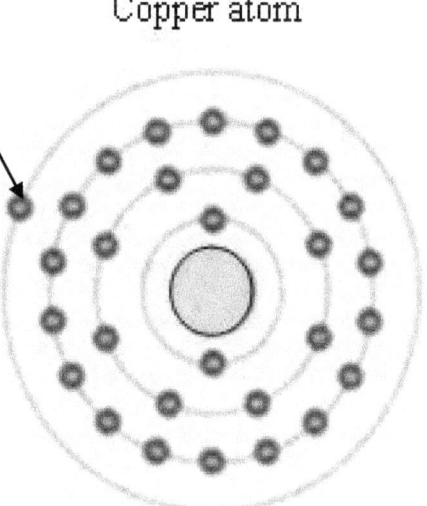

The Proton

Discovered by the great scientist **Ernest Rutherford** in the early years of the 20th century The Proton (*from the Greek word (first)*) Has a positive electric charge and a mass slightly less than that of a neutron. One or more protons are present in the nucleus of every atom. They are a necessary part of the nucleus. The number of protons in the nucleus is the defining property of an element, and is referred to as the atomic number. Each element has a unique number of protons, which means that each element is unique.

The Neutron

The Neutron is a subatomic particle, with no net electric charge and a mass slightly larger than that of a proton. Neutrons, are found in the nucleus of an atom, and in association with the proton are collectively referred to as nucleons. Their properties and interactions are described by nuclear physics.

Within the nucleus, protons and neutrons are bound together through the nuclear force, with neutrons being required for the stability of the atomic nucleus. Neutrons are produced copiously in nuclear fission and in nuclear fusion. Chemical elements within the stars are created through fission, fusion, and neutron capture processes.

The neutron was discovered in 1932 and is essential to the production of nuclear power. This applies to both to nuclear fission reactors, and nuclear fusion reactors; here on earth, and also in the stars, including our sun. Neutrons are used to induce many different types of nuclear transmutation, especially after the discovery of nuclear fission in 1938. It was quickly realized that, if a fission event produced neutrons, each of these neutrons might cause further fission events, etc., in a cascade known as a nuclear chain reaction.

Both Fission reactors and Fusion reactors are explained, and to understand the science it is necessary to understand what Protons and Neutrons are.

Individual neutrons, free of the nucleus, are effectively a form of ionizing radiation, and as such, are a biological hazard, depending upon dose. A small natural "neutron background" flux of free neutrons already exists on Earth, caused by cosmic ray showers, and by the natural radioactivity of spontaneously fissionable elements in the Earth's crust. Dedicated neutron sources like neutron generators, and research reactors produce free neutrons for use in irradiation and in neutron scattering experiments.

The History of Magnetism

Magnetism. Magnetism is a force generated in matter by the motion of electrons within its atoms. Magnetism and electricity represent different aspects of the force of electromagnetism. Magnetism was first discovered in the ancient world, when people noticed that a rock called lodestone, could attract towards it small pieces of the the iron based rock magnetite. The word magnet comes from the Greek term for lodestone, "magnítis líthos" which means a stone from the region of Magnesia. The Greek Philosopher Aristotle is attributed with the first of what could be called a scientific discussion on magnetism with the philosopher Thales of Miletus, who lived from about 625 BC to about 545 BC.

Electro Magnetism

An electromagnet is a type of magnet made from a large number of closely spaced turns of copper wire, wound around a core made from a ferromagnetic material such as iron. When and electric current is connected the magnetic core concentrates the magnetic flux and makes a strong. magnet.

The main advantage of an electromagnet over a permanent magnet is that the magnetic field can be quickly changed by controlling the amount of electric current in the windings. However, unlike a permanent magnet that needs no power, an electromagnet requires a continuous supply of current to maintain the magnetic field.

SECTION 9

Nuclear Power

Nuclear Fission

A Nuclear power plant is a type of power station that generates electricity using heat from nuclear reactions, and not from burning fossil fuels. The **Nuclear fuel** is a substance that is used in nuclear power stations to produce the heat which turns the turbines, which produce the electricity. In nuclear power stations this fuel is derived from two isotopes of uranium:-

E-238 and E-235. An isotope of a particular element such as uranium has a different number of neutrons in its nucleus than normal uranium. Nuclear fuel has the highest energy density of all practical fuel sources. Heat is created when nuclear fuel undergoes nuclear fission. Nuclear power does not emit gas pollutants. It is considered clean and environmentally friendly. It is a reliable, primary power source capable of producing electricity on the scale of the largest fossil fuel power stations. For example, the Hinkley Point nuclear power station in Somerset England has a generating capacity of 3.2 Giga Watts supplying 7% of UK electricity. Nuclear power stations, like those that burn coal, oil or natural gas, produce electricity by boiling water into steam. This steam drives a turbine, whose blades rotate an electric generator to create an electric current flow. It is exactly the same way that 200 years ago Michael Faraday generated an electric current, and he would easily recognise it today. The big difference is that nuclear power stations do not burn anything. Instead, they use the heat generated when uranium atoms are split open. This is called 'fission. The world's first commercial nuclear power station was at Calder Hall, Windscale, England. It began supplying 200 Mega watts of electricity to the UK National Grid, in 1956. There are now some 450 nuclear power stations around the word, and the number is increasing rapidly as China and other emerging major economies have adopted this method of generating electricity.

The Technical Details

There follows a more detailed explanation of the Nuclear fission process.

To make nuclear fuel from uranium ore requires first for the uranium to be extracted from the rock in which it is found. It is then enriched by bombarding it with Neutrons to create two types of uranium isotope: U-238 and isotope:-U-235.These two isotopes of uranium are made into uranium-oxide ceramic pellets; the equivalent to coal in a coal-fired power station. Some 50,000 cylindrical Zirconium Alloy tubes collected together in bundles are packed with the pellets. These now become the reactor Fuel Rods. Water under high pressure (*to prevent it from boiling*) is pumped in to the reactor core, and the fuel rods immersed. This is called a pressurised water reactor. PWR. Hinkley Point C represents the future of nuclear energy generation in the UK and a new type of pressurised water design promises higher output and greater efficiency with less waste. It is called a European pressurised water reactor 'EPR'. This is the primary coolant, and is pumped under high pressure into the reactor and then heated by the energy from nuclear fission. The Super heated water then flows to a steam generator and turned into superheated steam, which is then jetted onto the blades of a giant turbine. Which turns and generates an electric current. PWR designs have both uranium isotopes, and both have too many neutrons for them to exist for long in a stable state. The most salient **difference** is in how each isotope can be induced to split apart. Nuclear power plants obtain the heat needed to produce steam through a physical process. This process, called fission, entails splitting atoms of uranium in a nuclear reactor. The uranium fuel consists of the small, hard ceramic pellets already described above. A Nuclear fuel consists of the two types of uranium, U-238 and U-235. Most of the uranium is type U-238, but type U-235 splits or fissions much more easily. This is because in a U-235 nucleus, the number of protons and neutrons are very close to their maximum capacity. If one (or more) neutron is fired into the atom it will burst open shooting highly energetic neutrons into the more stable U-238 nuclei causing them also to break apart and shoot our neutrons like bullets. One fission triggers others, which triggers still more until there is a **chain reaction**. When that happens, in a split second the fission becomes self-sustaining; If it is not controlled there will be a mighty explosion --,an atomic bomb. To control this chain reaction special rods are inserted among the tubes holding the uranium fuel. These Control rods, are inserted or withdrawn to varying degrees, to slow or accelerate the fission rate of uranium. The rods are composed of chemical elements such as boron, silver, indium and cadmium all of which are capable of absorbing large numbers of neutrons without themselves fissioning. In effect mopping up the free neutrons and starving the chain reaction of fissile material. Water separates fuel tubes in the reactor, and the heat produced by fission turns this water into steam, which drives the turbine, as described above. Radioactive Uranium atoms fission by splitting into two smaller atoms together with a few neutrons, and the release of a large amount of heat energy. The new atoms are strontium-90, and strontium-89, both of which remain highly radioactive for possibly hundreds of years.

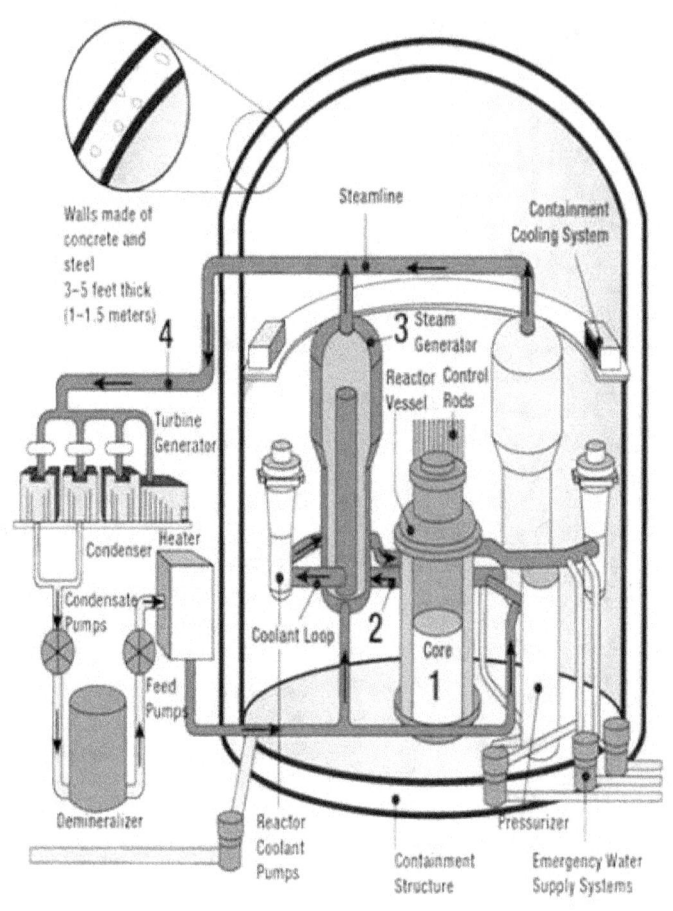

Walls made of
concrete and
steel
3-5 feet thick
(1-1.5 meters)

4

Steamline

Containment
Cooling System

Turbine
Generator

3 Steam
Generator

Reactor Control
Vessel Rods

Condenser Heater

Condensate
Pumps

Feed
Pumps

Coolant Loop 2

Core

1

Demineralizer

Reactor
Coolant
Pumps

Containment
Structure

Pressurizer

Emergency Water
Supply Systems

Illustration courtersy The U.S. Nuclear Regulatory Commission (NRC)

Nuclear Disasters

<u>Chernobyl</u> It was a catastrophic nuclear accident that occurred on 26 April 1986 at the Chernobyl Nuclear Power Plant in the city of Pripyat, then located in the Ukrainian Soviet Socialist Republic of the Soviet Union. An explosion and fire released large quantities of radioactive particles into the atmosphere, which spread over much of Western Europe. The Chernobyl disaster was the worst nuclear power plant accident in history in terms of cost and casualties. It is one of only two classified as a level 7 event (the maximum classification) on the International Nuclear Event Scale,. The struggle to contain the contamination and avert a greater catastrophe ultimately involved over 500,000 workers and cost an estimated 18 billion rubles. During the accident itself, 31 people died, and long-term effects such as cancers are still being investigated.

Fukushima This was a level 7 accident at the Fukushima, 'ONE' Nuclear Power Plant in Fukushima, Japan in March 2011. It was initiated primarily by a Tsunami, followed by the Tōhoku earthquake. Immediately after the earthquake, the active reactors automatically shut down their sustained fission reactions. However, the tsunami had destroyed the emergency generator cooling system for the reactors, causing reactor number 4 to overheat from its fuel rods. This led to three nuclear meltdowns and the release of radioactive material beginning on 12 March. Several hydrogen-air chemical explosions occurred between 12 March and 15 March. On 5th July 2012, the Fukushima Nuclear Accident Independent Investigation found that the causes of the accident had been foreseeable, and that the plant operator, Tokyo Electric Power Company (TEPCO), had failed to meet basic safety requirements such as risk assessment, preparing for containing collateral damage, and developing evacuation plans. On 12 October 2012, TEPCO admitted for the first time that it had failed to take necessary measures for fear of inviting lawsuits or protests against its nuclear plants. The Fukushima disaster is the largest nuclear disaster since the 1986 Chernobyl disaster and the second disaster to be given the Level 7 event classification of the International Nuclear Event Scale. Though there have been no fatalities linked to radiation due to the accident, the eventual number of cancer deaths, according to the Linear no-threshold theory of radiation safety, that will be caused by the accident is expected to be around 130-640 people in the years and decades ahead. The United Nations Scientific Committee on the Effects of Atomic Radiation and World Health Organization report that there will be no increase in miscarriages, stillbirths or physical and mental disorders in babies born after the accident. There are no clear plans for decommissioning the plant.

The Three Mile Island accident This was a partial nuclear meltdown that occurred on March 28, 1979, in reactor number 2 of Three Mile Island Nuclear Generating Station, Pennsylvania, United States. It was the most significant accident in U.S. commercial nuclear power plant history. The incident was classified 5 on the seven-point International Nuclear Event Scale: designated 'Accident With Wider Consequences'. The accident began with failures in the non-nuclear secondary system, followed by a stuck-open, pilot-operated relief valve in the primary system, which allowed large amounts of nuclear reactor coolant to escape. The mechanical failures were compounded by the initial failure of plant operators to recognize the situation as a 'loss-of-coolant accident'. In particular, a hidden indicator light led to an operator manually overriding the automatic emergency cooling system of the reactor because he mistakenly believed that there was too much coolant water present in the reactor for the, the steam pressure release. It was reported that the accident crystallized anti-nuclear safety concerns among activists and the general public, and resulted in new regulations for the nuclear industry. The partial meltdown resulted in the release of radioactive gases and radioactive iodine. There has been a substantial clean-up in the area, which ended in December 1993, at a cost of $1 billion. However, epidemiological studies analyzing the rate of cancer in and around the area since the accident, found no causal connection linking the accident with these cancers.

Nuclear Fussion

Nuclear Fussion Power

A Fusion reactor is all about containment. It is about keeping a superhot plasma of hydrogen isotopes trapped inside a magnetic field whilst nuclei inside it collide and fuse together releasing energy.

This is the most basic form of energy in the universe. It is what powers the sun (pictured) and all the other stars. It is produced by a nuclear reaction in which two atoms of the same light weight element, usually hydrogen, combine into a single atom of helium, the next heavier element after hydrogen in the table of elements, and in the process releasing a quantity of heat energy.

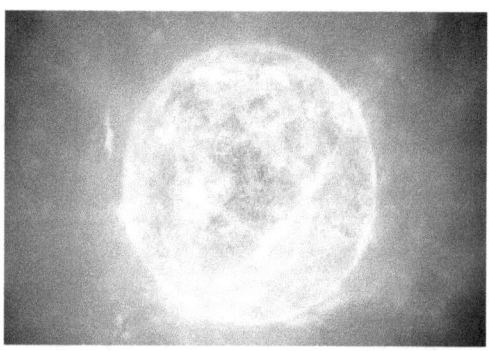

By detonating an atomic fission bomb encased in a cloud of Duterium/Tritium gas, scientists have successfully produced an uncontrolled fusion reaction to make a mighty explosion, hundreds of times greater than that of an atomic fission bomb. This is a **Hydrogen Fusion Bomb**. If the same amount of energy could be released gradually, as it is in a fission reaction, then what occurs in the sun, may well become the ultimate form of energy here on Earth. Seawater could become the basic ingredient of a clean inexhaustible fuel supply, potentially cheaper than any other fuel source on our planet. **Fusion power** is many times greater than **Fission power**, but frustratingly, creating a **controlled** fusion reaction has proven fiendishly difficult primarily because the two hydrogen derived nucleuses involved have the same positive charge. This means they will electrically repel each other, and simply refuse to merge. The rule is simple, "*like charges repel, unlike charges attract*", - and this repulsion reaction is very strong indeed. You can try it yourself with two magnets. Try forcing the two North pole ends together, and see how difficult it is. Nevertheless, the immense gravitational pressure and the 12 million degrees C° heat deep inside the sun's interior excites and accelerates the hydrogen nucleuses to the point where the mutual repulsion between them is overcome, and they merge together creating a nucleus of Helium and a lot of heat. When, in the far distant future, all the hydrogen in the sun has fused into helium the process will cease, the sun will switch off, and the earth will become a frozen lifeless ball of rock.

Here on earth an international consortium of some 15 nations has been striving for 30 years at colossal expense to build a prototype fusion reactor. Progress is being made, and it is hoped that one will be ready at a site in Southern France in about 10 years time.

The big problem is that although fusion takes place deep in the interior of the sun at around 12 million degrees centigrade the gravity and therefore pressure there is some one million times greater than that on our tiny earth. To produce fusion here 100 million degrees centigrade is required. Amazingly this has been achieved, but only for a fraction of a second.

We will make it work, and by the end of this century our great grand children could enjoy limitless amounts of virtually free fuel.

The Pros for nuclear fusion In many ways, fusion power seems like the perfect energy source. It's clean, it's inexpensive, and it uses seawater as its fuel source. It's the Holy Grail, it's the pot of gold at the end of the energy rainbow. It is safe and has no appreciable side effects. However, the opponents of all things nuclear say that the billions spent in research funding could be spent on renewable energy, and on helping the world's poor people instead.

SECTION 10

The Facts about climate change

As the 21st century comes to a close, our narrative approaches a closed door. Through it we glimpse an amazing yet today, unattainable world of new technology; one which will further change the lives of everyone on the planet. Just as the great Victorian pioneers of the industrial revolution Volta, Faraday, Siemens, Maxwell, Edison and others could, and did visualize Electric machines, Computers, Wireless communication, Flying machines, the Horseless carriage and low cost instant lighting for indoors and outdoors everywhere, (but strangely not the Internet).

These were highly intelligent men, who could and did foresee a new world ahead: a world that was unattainable to them, but is common place to us; so we too can try to see applications which are today embryonic, and just notes at a scientific conference, but will one day be as normal to our great grandchildren as colour television is today. If we succeed, and we must succeed, in halting the world's shrinking land mass, then, to paraphrase the immortal words of Sir Winston Churchill, our civilization will continue and the life of the world may move forward, "into broad, sunlit uplands. If we fail, then the whole world, including the United States, including all the developed and developing countries could sink into the abyss of a new Dark Age, made more sinister, and perhaps more protracted, by border conflicts, mass migrations, fanaticism, terrorism and perverted science".

If sea levels continue to rise, then by 2150, Island nations such as Fiji and the Maldives will disappear under the Ocean, along with hundreds of other low lying islands around the world. Shorelines everywhere will move inland as sea levels rise, and many of the world's great cities will need high-tide protection barriers. Elsewhere, the Sahara desert will continue its creep southward, making The Republic of Mali habitable only to wandering Tuareg nomads. It is called 'Global Warming', and caused by vast invisible clouds of chlorofluorocarbon gas

(CFC's) being pumped into the atmosphere as a bi-product of burning carbon fuels:-- Coal, Oil and Natural Gas. These gases, especially Carbon Dioxide, absorb Infrared Radiation from the Sun which then makes the gas warmer, and this makes the atmosphere warmer. They are known as Green House Gases.

The greenhouse effect is a natural process that is essential to life on earth, but the build-up of greenhouse gases such as carbon dioxide, methane and nitrous oxide in the atmosphere is amplifying the natural effect and destabilizing the balance. Greenhouse gas slows down the heat escaping from our planet, and just like a cosy Duvet, makes everything underneath it warmer. If nothing is done to address the natural balance, then scientists are warning that the build-up of Greenhouse gas emissions risks raising average global temperature by 4°C towards the end of this century. Without doubt this would be a catastrophe -- and the blame lies squarely with the fuel we are using to support our rapidly growing populations and economies.

Human activity is responsible for almost all of the increase in greenhouse gas in the atmosphere. Burning Fossil fuels: -- Coal, Gas and Oil, between them generate millions of tons of waste products annually in the form of gas, primarily: Carbon dioxide, Methane, Nitrous oxide and Ozone; Labelled:- Chlorofluorocarbon organic compounds or (CFC's).

These are the Green House gases.

Slowly the seasons are shifting, November's weather is like that which we traditionally experienced in October. December days, are now mostly warm, wet and windy. Increasingly we hear a Woodpecker drumming in the woods. It is at least a month earlier than what the book says. Over the past few years most people have become acutely aware of changes in local weather patterns, with British winters becoming noticeably milder, with snow in the lowland areas, a rarity that most five year old children have never seen. In central Europe and in North America torrential rain storms with severe flooding seem to be an annual event, experienced more frequently than most people can remember.

After debating the issue for at least thirty years most experts finally accept the warming trend in our climate is of significance. With technological advances including satellites, and powerful computers, scientists have collected data from all over the world. This includes wind, and ocean currents, the Jet Stream and the chemical composition of the atmosphere from all around the globe. Sophisticated computer models have been created which provide compelling evidence for rapid climate change; confirmation perhaps of the experience of ordinary people.

Average global temperatures have now been recorded as the highest ever. This is causing the melting of Arctic sea-ice, the disappearance of Glaciers in Switzerland, and severe annual flooding from monsoon-like rainfall in Britain and Central Europe. Elsewhere, drought is having a devastating impact on agriculture, whilst the Sahara desert is moving steadily southward. Indeed as the charts below clearly indicate, average global temperatures have been rising for over a century, and are actually speeding up over the last few years. The Earth's atmosphere is warming, faster than it probably ever has. In many cases weather patterns, climates and natural environments are changing even quicker than predicted. If present rates continue unchecked, the Polar Bear will have become extinct in the wild, by 2070, and an ice-free North Pole will be a summertime cruise-ship destination, as early as 2025.

Climate Change: The Hard Truth

The following report is from the American 'National Aeronautics and Space Administration' NASA, in January 2017.

In 2016 the Earth's surface temperatures were their warmest since modern record keeping began in 1880. According to independent analyses, globally-averaged temperatures in 2016 were 0.99 degrees Celsius warmer than the mid-20th century. This makes 2016 the third year in a row to set a new record for global surface temperatures, and continues a long-term warming trend. However, the scientists say that because weather station locations and measurement practices change over time, there are some uncertainties in the exact interpretation of any specific year, although, even taking this into account, NASA insists that 2016 was the warmest year yet. Remarkably this is the third record year in a row, and the long-term warming trend is clear, the planet's average surface temperature has risen about 1.1 degrees Celsius) since the late 19th century. A change driven largely by increased carbon dioxide and other human-made emissions detected in the atmosphere. Most of this warming occurred in the past 35 years, with 16 of the 17 warmest years on record occurring since 2001. Phenomena such as **El Niño** and **La Niña**, which warm or cool the upper tropical Pacific Ocean, cause short-term variations in global average temperature, and a warming. The **El Niño** event which was in effect during 2015, increased the annual global temperature anomaly for 2016 by 0.12 degrees Celsius. NASA's analyses incorporated surface temperature measurements from 6,300 weather stations, ship- and buoy-based observations of sea surface temperatures, and temperature measurements from Antarctic research stations. These raw measurements were analyzed using an algorithm that considers the varied spacing of temperature stations around the globe and urban heating effects that could skew the conclusions. With results such as these there can surely no longer be any doubt that increased levels of greenhouse gases are causing the Earth to warm up in response.

Ice cores drawn from Greenland, Antarctica, and mountain glaciers show that Greenland lost 150 to 250 cubic km (*36 to 60 cubic miles*)   of ice per year between 2002 and 2006, while Antarctica lost about 152 cubic km (*36 cubic miles*) of ice between 2002 and 2005. Both the extent and thickness of Arctic sea ice has declined rapidly over the last several decades. Glaciers are retreating almost everywhere around the world — including in the Alps, Himalayas, Andes, Rockies, Alaska and Africa. The earth's climate is responding to changes in greenhouse gas levels. Global sea level rose about 17cm (*6.7 inches*) in the last century. All three major global surface temperature reconstructions show that Earth is definitely getting warmer.

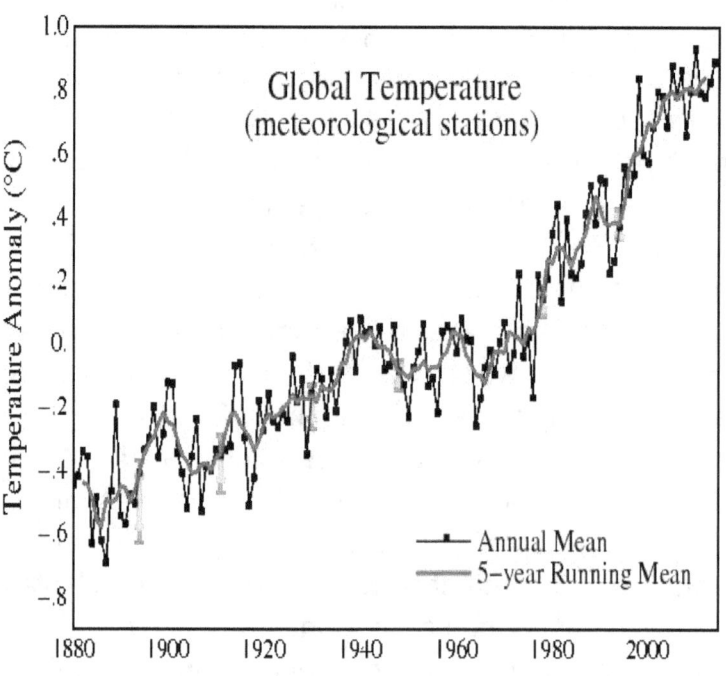

180

The Green House Gases

This article appeared in the London Times Newspaper in December 2016.

It was 160 years ago that an American scientist demonstrated how gases in the Earth's atmosphere interact with the sun's rays and trap heat on earth. Of all the gases tested it was carbonic acid gas, known today as carbon dioxide that trapped the most heat. This discovery of the greenhouse effect was made public in 1856 about three years before the British physicist **John Tyndall** presented the same findings in London. Tyndall got all the credit though, because of one simple fact; the scientist who made the original breakthrough was a woman, **Eunice Foote**. Her work was demonstrated to a science conference in New York State in 1856, but it was presented on her behalf by Joseph Henry, a professor at the renowned Smithsonian Institution. He described how Foote took glass jars containing water vapor, carbon dioxide and air. She then heated them up in the sun and compared the results. "The highest effect of the sun's rays I have found to be in carbonic acid gas. The receiver containing the gas itself became much heated very sensibly more so than the other -- and on being removed it was many times as long in cooling."

Foote also speculated whether even modest changes in the amount of carbon dioxide in the air could lead to significant changes in global temperatures. This was the classic explanation of the greenhouse effect, but Foote's work was not published in a science journal, and we know about it only from a journalistic account published in an annual review of worldwide scientific achievements for the year 1856. It appears Foote could not reach a wider audience because she was a **woman**. Tyndall appeared unaware of this earlier Discovery.

However his publications were more extensive, including accurate measurements of how much different gases absorbed radiant heat from the Sun. His work has become the Milestone in research on the greenhouse effect, and Foote's name is unfairly consigned to a footnote in history.

The following statistics were reported on Wikipedia in 2016

Levels of Carbon dioxide in Earth's atmosphere have varied, ranging from as high as 7,000 ppm during the Cambrian period about 500 million years ago to as low as 180 ppm during the Quaternary glaciation of the last two million years.

Carbon dioxide is an integral part of the carbon cycle, a bio-geochemical cycle in which carbon is exchanged between the Earth's oceans, soil, rocks and biosphere. The present biosphere of Earth is dependent on atmospheric CO_2 for its existence. Plants use solar energy to synthesise carbohydrate from atmospheric carbon dioxide and water by photosynthesis. Carbohydrate derived from consumption of plants as food is the primary source of energy and carbon compounds in almost all other organisms.

The current episode of global warming is attributed to increasing emissions of CO_2 and other greenhouse gases into Earth's atmosphere, and stems from the pioneering work of John Tyndall and Eunice Foote at the start of the Industrial Revolution. The global annual mean concentration of CO_2 in the atmosphere has increased by more than 40% since then, from 280 ppm, the level it had for the last 10,000 years leading up to the mid-18th century, to 399 ppm as of 2015 The present concentration is the highest in at least the past 800,000 years and likely the highest in the past 20 million years. The increase has been attributed to anthropogenic sources, particularly the burning of fossil fuels and deforestation The daily average concentration of atmospheric CO_2 at Mauna Loa Observatory first exceeded 400 ppm on 10 May 2013 It is currently rising at a rate of approximately 2 ppm/year and accelerating. An estimated 30–40% of the CO_2 released by humans into the atmosphere dissolves into oceans, rivers and lakes, which contributes to ocean acidification.

So now we know that Climate change is primarily caused by the accumulation of greenhouse gases, or heat-trapping gases, in the atmosphere. We also know that Greenhouse gases are certain molecules in the air that have the ability to trap heat in the Earth's atmosphere. These are called Chlorofluorocarbons (**CFCs**), Some greenhouse gases, like carbon dioxide (CO_2) and methane (CH_4), occur naturally and play an important role in Earth's climate. If they didn't exist, the planet would be a much colder place, and perhaps uninhabitable. However, as we have seen, some greenhouse gases are entirely man-made and are products of human activities, such as burning fossil fuels (*i.e. coal, oil and natural gas*) for Heat energy and transport. They emit carbon dioxide molecules CO_2 and other greenhouse gases. These add to the natural ones in the atmosphere building up to cause an overall increase in the warming of the planet. All greenhouse gases are created equal in terms of contributing to climate change. Their impact varies according to how long they remain in the atmosphere and how efficient they are at trapping heat. For example, methane, the main component in natural gas, remains in the atmosphere for a shorter time than CO_2, but is far more efficient at trapping heat, making it a more potent greenhouse gas. Nitrous oxide is less abundant than methane, but even more efficient at trapping heat, and it stays in the atmosphere for a long time.

Understanding the impacts and sources of the main greenhouse gases can help inform strategies designed to reduce them.

The industrialised world produces far more greenhouse gas than any other man-made source, with CO_2 (*carbon dioxide*) the one you hear people talk about the most. In the USA 30% of green house gas emission is from power stations generating electricity. Although the developed economies of the world are by far the biggest emitters of green house gas, countries such as Brazil, Indonesia and India, play a critical role by deforestation. Forests absorb carbon dioxide, and chopping them down on a huge scale removes an important sponge of these gases

From the charts below we can see that in the USA, among the Economic Sectors, the totals of Electricity generation, and transportation are responsible for 56% of all Greenhouse gas emissions. These statistics are similar throughout the developed world, and suggest that if we could generate our electricity by non polluting methods, and change the fuel we use for transport to electricity, then, just to emphasise, we have a good chance to halt global warming.

Large contributors to man-made climate change in 2014 were coal fired power stations which accounted for 39% of electricity generated in the United States. In that year, the electricity sector was the largest source of U.S. greenhouse gas emissions, accounting for about 30 percent of the U.S. total.

UK greenhouse gas emissions from electricity generation account for around a quarter of the UK total. Between 2009 and 2014 power sector emissions declined on average by 4% per annum, primarily as a consequence of moving away from burning coal. In the developing world coal remains the primary source of fuel for electricity generation, and countries such as China and India are now among the largest sources of greenhouse gasses.

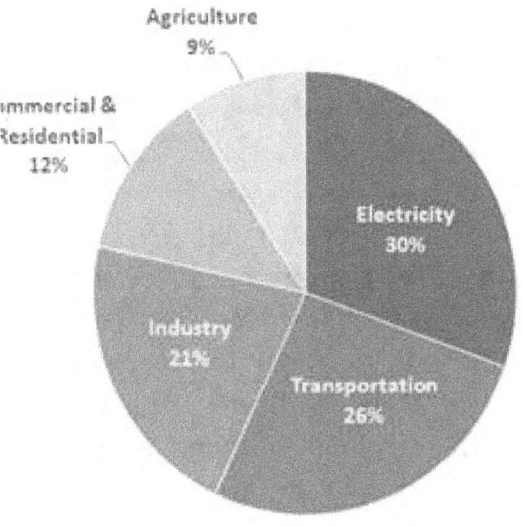

The Case against Climate Change

Despite all the scientific data collected from around the world there are still respected scientific voices expressing doubt that human- caused climate warming is indeed happening.

Dr. **S. Fred Singer** is a distinguished astrophysicist who has taken strong stance against the evidence for climate change. In this book,'**Hot Talk Cold Science'** published in 1997, Dr. Singer explores the inaccuracies in historical climate data, the limitations of attempting to model climate on computers, and the impact of solar variability and its impact on climate. Also, what factor could possibly mitigate any human impacts on world climate. Dr. Singer believes that global warming scenarios depicted in the media have no scientific basis. His argument that there has been no global warming during the past twenty years has, and even if there has, he finds that many aspects of any global warming, such as a longer growing season for food and a reduced need to use fossil fuels for heating, would actually have a positive impact on the human race. However, from the highest quality temperature data in both the surface record and satellite measurements, it is quite clear that global warming has continued unabated during the 21'st century. Rather than embark on economically destructive policies to solve a problem that to the best of our knowledge does not exist, Singer urges policymakers to adopt a "no regrets" policy of continued research and unimpeded economic growth. We would then have more scientific knowledge, technology, and economic resources with which to confront climate warming, if we ever discover that it is occurring and poses a real threat. But prematurely mandating severe reductions of greenhouse gas emissions would make us—and developing countries, especially—poorer and less able to cope with any future problems. Human activity is adding carbon dioxide to Earth's atmosphere. While the carbon dioxide concentration varies naturally over long periods of time, and measurements show that it is now far higher than it has been for at least 800,000 years. Indeed, it is 40% higher than 150 years ago. There is absolutely no scientific dispute about the measured rise in the carbon dioxide concentration, and there is also no doubt that the rise is a result of fossil fuel use, which releases carbon dioxide with a different isotopic signature than that of other carbon dioxide in the atmosphere.

Is everything bad about Global Warming?

Given the incessant talk about the purported catastrophes that global warming might cause—severe storms, severe heat-waves, sea level rise, the world's deserts getting bigger; it sounds strange to hear about benefits from global warming. Nevertheless, the scientific literature does support the view that increases in CO_2 concentration and global temperatures, could actually improve human well-being in the high latitude temperate zones. These benefits include a CO_2-enriched biosphere more conducive to plant growth, longer frost-free growing seasons, and greater water efficiency for plants.

However, despite the doubters, and despite the rejection by some scientists and politicians, for an interpretation of measurements, which suggest that earth's climate is indeed changing. The concensus by everyone else is that it is. Weather patterns have changed, Spring in the Northern hemisphere comes earlier than it did. Alarming weather conditions that seem to have become a regular occurrence in recent years with devastating , hurricanes, drought, fire and floods in regions where they were previously described as 'once, in- a 'life-time' events. The only explanation that seems to fit is that this change in climate is caused by man-kind. The gases that are blamed for causing the earth's temperature to increase are called Chlorofluorocarbons (**CFCs**)n and their increase is overwhelmingly attributed to burning fossil fuels, as explained earlier in this section. Despite the doubters, we must switch to more non polluting fuels when generating electricity, or heating our homes and offices. That basically means using more electrically generated power and heat, and to generate these with low cost climate friendly clean electrical machines. This includes our surface transport vehicles.

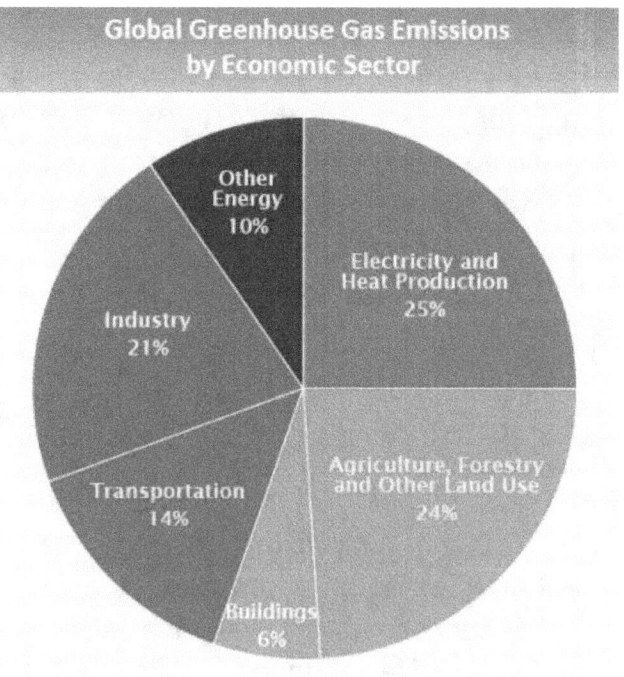

Global Carbon Emissions from Fossil Fuels, 1900-2011

management activities.

Agriculture, Forestry, and Other Land Use (24% of 2010 global greenhouse gas emissions. Greenhouse gas emissions from this sector come mostly from agriculture (cultivation of crops and livestock.

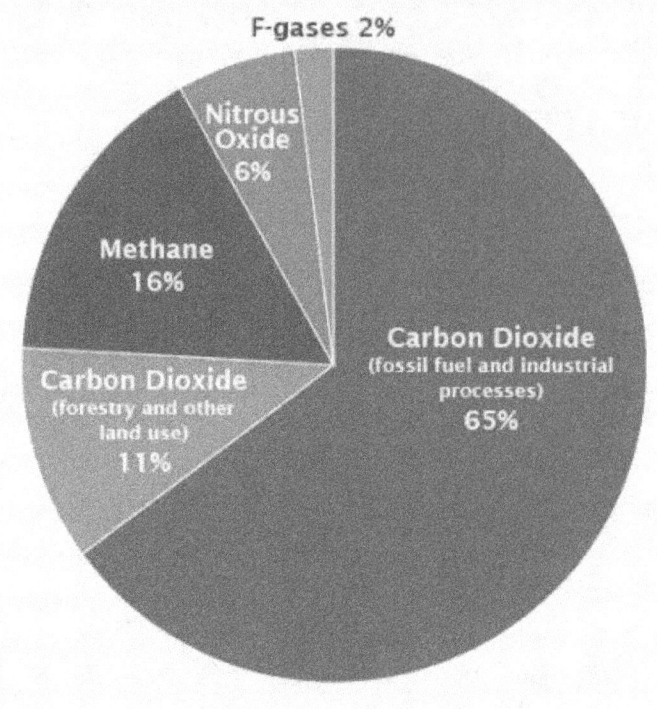

Global Greenhouse Gas Emissions by Gas

F-gases 2%

Nitrous Oxide 6%

Methane 16%

Carbon Dioxide (forestry and other land use) 11%

Carbon Dioxide (fossil fuel and industrial processes) 65%

SECTION 11

The Great 19th century Pioneers of Electrical Science

SECTION 11

The Great 19th century Pioneers of Electrical Science

The Battery

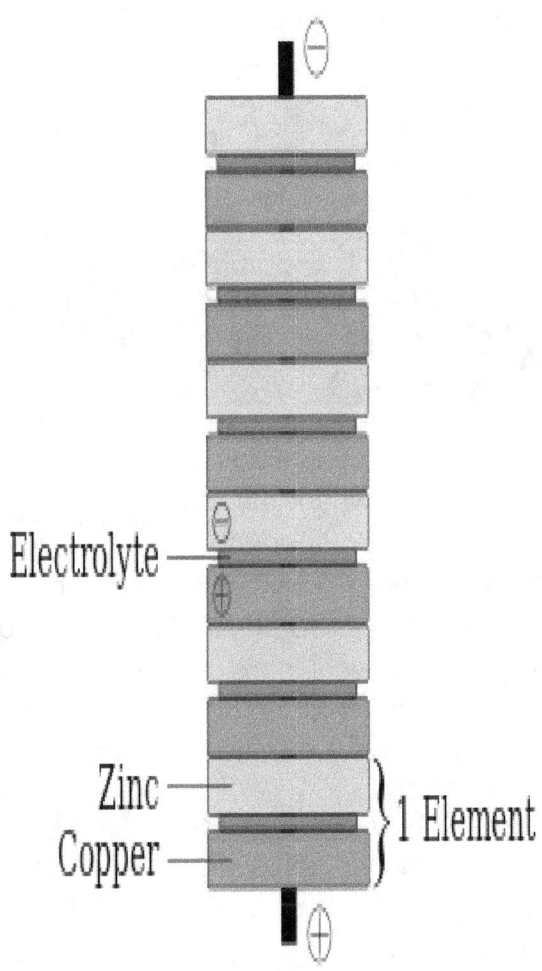

Electrolyte

Zinc
Copper

} 1 Element

Alessandro Volta

The ability to generate a continuous electric current. took a giant step forward 1799 when the eminent Italian scientist **Alessandro Volta**, (*in whose honour the unit of electrical force, the Volt, was named*), built a primitive battery, which he called a Voltaic Pile'. It consisted of two electrodes: one made of zinc, the other of copper which were separated by paper soaked in brine, which he *later changed to dilute sulphuric acid, and called* The Electrolyte.

The chemical reaction between the two electrodes caused the zinc electrode to become negative and the copper electrode positive. By stacking groups of these cells in a pile and connecting the two end terminals together, a constant electric current was seen to flow, Researchers into the electric phenomena now had a more reliable source of electrical energy than the electrostatic machines currently being used. A massive impetus was triggered into the understanding and science of electricity. After demonstrating his 'Pile' in Paris and then to the Royal Institute in London, Volta convincingly proved that electricity could be generated chemically. This contradicted the long held belief that it could only be generated inside one of god's living creatures. His invention ignited a huge response and controversy reigned in the religious and the scientific worlds. Alessandro Volta, a devout Christian, was under real threat of excommunication from the catholic church. Fortunately, he managed to satisfy most of his critics by affirming his strong Christian beliefs, and declaring that his discovery was the work of god, and waivered his patent rights. Incidentally Volta's pile of voltaic cells reminded people of a row of canon guns, so they called the assembly a Battery..

Hans Christian Ørsted

Danish Chemist and Physicist.

Born: 14 August 1777Died: 9 March 1851 aged 73

Credited with the discovery of electro-magnetism.

Oersted launched a new epoch in science when he discovered that electricity and magnetism are linked. He showed by experiment that an electric current flowing through a wire could move a nearby magnet. The discovery of electromagnetism set the stage for the development of modern electro-magnetic theory.

Michael Faraday

Born: 22 Sept 1791

Died: 25 August 1867

Michael Faraday FRS was an English scientist who made a major contribution to the study of of electromagnetism. His main discoveries include the principles underlying electromagnetic induction, and Electrolysis.
Although Faraday received little formal education, he was one of the most influential scientists in history. It was by his research on the magnetic field around a conductor carrying a direct current that Faraday established the basis for the concept of the electromagnetic field in physics. Faraday also established that magnetism could affect rays of light and that there was an underlying relationship between the two phenomena. He similarly discovered the principles of electromagnetic induction and diamagnetism, and the laws of electrolysis. His inventions of electromagnetic rotary devices formed the foundation of electric motor technology, and it was largely due to his efforts that electricity became practical for

use in technology. Faraday ultimately became the first and foremost Fullerian Professor of Chemistry at the Royal Institution, a lifetime position. He was an excellent experimentalist who conveyed his ideas in clear and simple language at very popular talks and demonstrations.

Faraday's Electric Generator

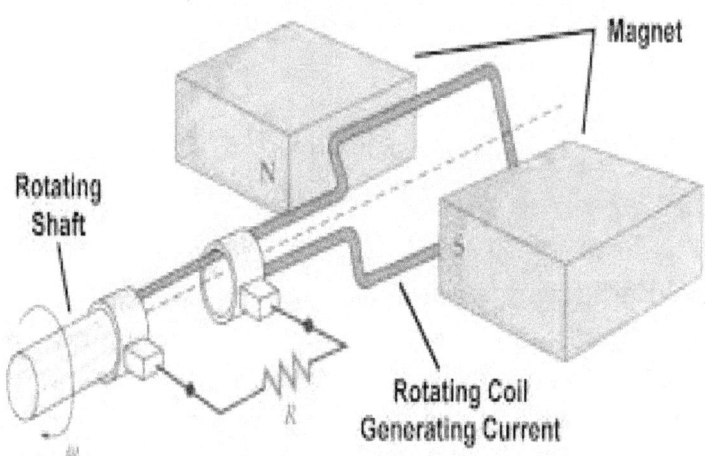

Faraday's Electric Generator

Michael Faradays mathematical abilities, did not extend as far as trigonometry and were limited to the simplest algebra. James Clerk Maxwell took the work of Faraday and others and summarized it in a set of equations which is accepted as the basis of all modern theories of electromagnetic phenomena. On Faraday's uses of lines of force, Maxwell wrote that they show Faraday "to have been in reality a mathematician of a very high order – one from whom the mathematicians of the future may derive valuable and fertile methods." The SI unit of capacitance is named in his honour: **the Farad**. Albert Einstein kept a picture of Faraday on his study wall, alongside pictures of Isaac Newton and James Clerk Maxwell. Physicist Ernest Rutherford stated, "When we consider the magnitude and extent of his discoveries and their influence on the progress of science and of industry, there is no honour too great to pay to the memory of Faraday, one of the greatest scientific discoverers of all time."

Thomas Alva Edison

Born 18 Feb. 1847 Died 18 Oct 1931

Thomas Alva Edison **was an** American inventor and businessman, who was probably America's greatest inventor, with 1,093 US patents in his name. in addition to as many patents in the United Kingdom, France, and Germany.

The electric light, is one of the everyday conveniences that most affects our lives. It was the first commercially practical incandescent light in the world. Many of the devices that he played a key role in inventing, and bringing to market, greatly influences the world we inhabit today. These include the phonograph, the motion picture camera, and the electric light bulb.

He was the first technologist to apply the principles of mass production and large-scale teamwork to the process of invention, and development in a laboratory environment.

Edison developed a system of electric-power generation and distribution to homes, businesses, and factories – a crucial development in the modern industrialized world.

Nicolai Tesla

July 1856 – 7 January 1943) was a Serbian-American inventor, electrical engineer, mechanical engineer, physicist, and futurist who is best known for his contributions to the design of the modern alternating current (AC)

Nikola **Tesla** was a creative electrical engineer. Best known for designing the alternating-current (AC) electric system, which is still the predominant electrical system used across the world today. He sold several patent rights, to Thomas Edison, also to The Westinghouse company for his patents in AC **Other** Alternating current projects include , high-voltage, high-frequency power experiments.

Tesla had a rather heated and highly publicized feud with Thomas Edison over whose electrical system would power the world — **Tesla's** alternating-current (AC) system or **Edison's** rival direct-current (DC) system. The two feuding geniuses waged a "War of Currents" in the 1880s. The winner was of course Tesla's AC system.

James Clerk Maxwell 1831-1879

James Clerk Maxwell

James Clerk Maxwell was probably the most influential scientists of his time. Albert Einstein acknowledged that the origins of the special theory of relativity lay in Clerk Maxwell's theories, saying "The work of James Clerk Maxwell changed the world forever". Clerk Maxwell's research into electromagnetic radiation led to Albert Einstein's Theory of Relativity.

James Clerk Maxwell was born in Edinburgh in 1831. He attended school in the city and later studied at the Universities of Edinburgh and Cambridge. He was an intensely curious child, writing his first scientific paper at the age of 14. At 25 he became Professor of Physics at Aberdeen University's Marischal College. Whilst the developed world was in the middle of its fascination with electricity and magnetism a momentous discovery was made, without anyone realising it. In 1865 whilst building a mathematical model of Michael Faraday's empirical connection between electricity and magnetism, Clerk Maxwell was able to demonstrate mathematically that electric and magnetic waves were part of the same phenomena as light waves. Michael Faraday had explained electromagnetic induction using a concept he called lines of force. However, scientists at the time wanted a mathematical model, and James Clerk Maxwell, who was a highly respected Mathematician was able to produce it. Maxwell demonstrated that electric and magnetic fields travel through space as waves moving at the speed of light. Maxwell proposed that light itself is an undulation in the same medium that causes electric and magnetic phenomena. The unification of light and electrical phenomena led to the prediction of the existence of other waves, later to be called Radio Waves. Some 22 years later In1887, German scientist Heinrich Hertz demonstrated the reality of Maxwell's electromagnetic waves by experimentally generating radio waves in his laboratory. The age of Radio, Television and the i-phone had begun.

Today, Maxwell's Field Equations form the basis of our modern understanding of electro magnetismJames Clerk Maxwell is one of the most influential scientists of all time. Albert Einstein acknowledged that the origins of the special theory of relativity lay in Clerk Maxwell's theories, saying "The work of James Clerk Maxwell changed the world forever".

Maxwells Field equations demonstrate mathematically that electric and magnetic waves are part of the same phenomena as light waves, and predicted radio waves

Maxwell's equation for electric fields **E** and magnetic fields **H** are:

$$\text{div } \mathbf{D} = \rho_v$$

$$\text{curl } \mathbf{E} = -\frac{\partial \mathbf{B}}{\partial t} = -\mu_o \mu_r \frac{\partial \mathbf{H}}{\partial t}$$

$$\text{curl } \mathbf{H} = \mathbf{J} + \frac{\partial \mathbf{D}}{\partial t} = \epsilon_o \epsilon_r \frac{\partial \mathbf{E}}{\partial t}$$

$$\text{div } \mathbf{B} = 0$$

where μ_o is the permeability of free space, μ_r the permeability of the material, **D** electric displacement, **J** current density, ρ_v volume charge density, ϵ_o and ϵ_r permittivity of free space and permittivity of the material respectively.

Werner von Siemens

Inventor and Industrialist

In 1847 Werner von Siemens and Johann Halske started a company to produce an electric telegraph. Their invention used a needle to point to a sequence of letters, instead of using the Morse code.

In 1848, the company built the first long-distance telegraph line in Europe; 500 km from Berlin to Frankfurt am Main. In 1850, the company opened a London branch headed by, Carl Wilhelm Siemens, (later Knighted to, Sir William Siemens). The company was involved in building long distance telegraph networks in Russia. In 1855, a company branch headed by another brother, Carl Heinrich von Siemens, was opened in St Petersburg, Russia. In 1867, Siemens completed the monumental London to Calcutta telegraph line, an amazing engineering project at the time.

The Dynamo

Because of Michael Faraday's discovery of machines using electro-magnetic induction to produce an electric motor or an electric generator, the use of permanent iron magnets had to be used to produce the essential magnetic field. Because the power of a Generator is directly dependent upon the power of the magnetic field the size of electric machines was severely limited. The stronger the magnetic field the more powerful the machine, and a permanent magnet was simply not powerful enough.

"We have a similar impasse today with batteries".

Because of Faraday's discovery the mania for more and more powerful electric machines, brought the evolution of the battery to almost a stop. Then in 1867 life for the battery got even worse. Werner von Siemens described a device which could generate its own magnetic field, and at almost the same time Mr. Charles Wheatstone described another. -- the log jam was broken: without the need for an iron permanent magnet. an electric machine could be scaled up to almost any size.

The device became known as a 'Dynamo' which is a Greek word Dynamis, meaning power,

Both inventors were awarded a patent, but Siemens became the first company to manufacture dynamo devices. The innovation of a stator field, rather than a permanent magnet was a giant technological leap forward enabling motors and generators to be built capable of powering the largest industrial applications. Before long Siemens were delivering Electric Railway Trains, Power station turbines, and the first electric arc furnaces for the production of steel and other materials. The company's first electric locomotive, was built in 1879 and in

1881 a Siemens AC Alternator driven by a watermill was used to power the world's first electric street lighting in the town of Godalming, United Kingdom. The company has continued to grow and branch out around the world, today building giant on-shore and offshore wind turbines. Siemens AG is one of the largest electro-technology firms in the world, and the von Siemens family still owns 6% of the company.

Siemens Power station Turbine

Beyond The Beginning

A new Frontier

Beyond The Beginning

The great changeover is under way, and this book has described the technology, and the products now beginning to be produced. In particular the '**Super Battery**, which really does exist**,** and when it moves out of the laboratory and in to mass production, will give a driving range comparable to, or better than, the average diesel or petrol engine.

Meanwhile, the **ultimate** battery, that can provide energy to power your car or portable device for its entire lifetime, is still only a dream. But, it is not an impossible dream, and it could come true

<u>**Now consider this**</u>: In 1977, two spacecraft began a mission that has redefined our knowledge of the Solar System. Remarkably, after 41 years both Voyager 1 and Voyager 2 spacecraft are still working, and sending very weak signals back to earth. Voyager 1 left the solar system in 2013 and is entering interstellar space, travelling at 26,900 miles per hour. It is now over 20 billion kilometres (12 billion miles) away, and it takes a radio signal, travelling at 186,000 miles per second (*the speed of*

light) 38 hours to reach earth and be detected by Nasa's giant satellite dishes.

The electricity to power the spacecraft's instruments and radio transmitter comes from a **radio-isotope thermoelectric generator,** or RTG. This uses radioactive materials (such as plutonium) to generate heat which can then be converted into an electric current to charge a battery. In the case of Voyager 1 it originally produced 155 watts, and after 41 years the battery can still provide a much reduced electric current. The only reason that we cannot have a battery like this today is that the highly radioactive material would kill a motorist. However, a safe non-radioactive alternative technology already exists, if only we could get it to work. Nuclear fusion could be the answer, and a nuclear fusion battery could easily last for 45 years. Although Hydrogen nuclear fusion power generation is still some years away, scientists are making encouraging progress and an International collaboration has funded a test power station that is now being built in France. It is scheduled for operation around 2030. Grid size power stations are clearly on the way. Sadly this encouraging progress is still a long way from producing a fusion battery. The main difference, is that *a low energy nuclear reaction (LENR) , or* cold fusion process, is necessary to make a battery, and this has never been positively demonstrated, despite efforts by scientists around the world. Even the Space Agency NASA's attempt to produce low energy cold fusion resulted in more input power required than power produced. The key to cold fusion's environmental cleanliness and safety seems to be the slow-moving atomic particles (neutrons). Whereas standard nuclear fission and nuclear fusion both create fast moving high energy neutrons, which create massive damage when they smash into the nuclei of other atoms, cold fusion LENR produces particles with an energy less than a millionth of this, and do not generate ionizing radiation or radioactive waste. It is because of this sedate gentility that LENR lends itself very well to generating electric current to charge batteries of every size.

The cold fusion dream lives on:

NASA and others are pressing on towards the goal of a cheap, clean, low-energy nuclear reaction, cold fusion technology that could eventually see cars, planes, and homes powered by small, safe nuclear reactors. NASA dreams of one day putting a cold fusion reactor in every home, car, plane, and iphone. The device will come with an installed battery which lasts for the equipment's lifetime.

Today, the insurmountable barrier to cold fusion is that the energy required to get a reaction is much greater than what is produced. However, there are no equations, measurements or theories, declaring that it is impossible. Cold Fusion is possible, and it is realistic to predict that by around 2070 the dreams of these great pioneers will be realised.

INDEX

	Glossary -- Words and names found in the book, which may not be familiar to many readers.
Lithium	Silver-white element the lightest metal known.
Solar Cell	A Solar Cell converts light into electricity.
Battery	A Chemical device for storing electrical energy.
Green House Gas	A gas that absorbs infrared radiation and radiates heat.
EV	An Electric Vehicle
Alessandra Volta	Italian scientist who in 1799 invented the battery
Michael Faraday	Inventor of the electric motor and generator. 1810.
Karl Benz	German inventor of the internal combustion engine 1886.
Albert Einstein	world famous scientist.
Negative Ion	An atom with a an extra electron.
Positive ion	An atom with a missing electron.
James Dyson	Inventor and entrepreneur.
Professor Geim	co inventor of Graphene
Kostyar Novoselov	co inventor of Graphene
Graphene	The world's first 2D material with amazing attributes.
2D Materials	2D materials have length, breadth and 1 atom thick depth.
Lithium Sulfur	(Li–S battery) is notable for its high specific energy.
Lithium ion	(Li–ion battery) Rugged standard EV battery in use today.
Lithium Air	(Lithium Air) This could be the 'Super' battery for EV's.
Road Charging	A charge to motorists per Km travelled.
Petroleum	Gasolene
US Gallon	3.785 litres
UK Gallon	4.546 litres
Road Train	Commercial vehicles coupled as together by wi-fi or chain
Primary battery	None rechargeable, Chemical storage
secondary "	Rechargeable, Chemical storage
Close coupling	Magnetic induction between two separated circuits.
SMOG	Chemical fog caused by fossil fuels
Photon	A particle of light.
Speed of Light	three hundred million metres a second.
Laser	Light amplification by the stimulated emission of radiation

Sources of Information

Reports on motor transport

UK Department of Transport UK Automobile Associations, (The AA) UK UK Road Transport Authority'

Kevin Nicholson, head of tax at PwC, said "the government would need to move quickly to reverse a decline in fuel duty. Report from: The UK Treasury's official forecaster, 2017 reports ffrom The Office for Budget Responsibility.

Bloomberg business reports

The Financial Times of, 2017, technology reports

Report.July 2017 The UK Government announce another tranche of £246 million for the 'Michael Faraday battery development initiative'

An important Study on Cycle-Life Prediction of Lithium-Ion, Batteries for Electric Vehicles Published in the 'Journal of Electrochemical Science. (www.electrochemsci.org) is a valuable contribution to this subject. Authors: Minghui Hu , Jianwen Wang, Chunyun Fu

Reports from the International Energy Agency IEA). on countries that have announced plans to switch to electric cars by the the mid 2020's, and replace fossil fuels. These include:- Austria, Denmark, Ireland, Japan, India, the Netherlands, Portugal, Korea and Spain

Public announcements from American manufacturing companies about their planned launch of electric trucks:

Reports from The United States Environmental Air quality bureau on California Roads.

The American Automobile Association Road casualty report. Battery technology report from The Brookings Institution

. The Sakti3, a start-up, launched from the University of Michigan by Professor Ann Marie Sastry. The Sakti3 team has published over 100 papers concerning super battery technology, and is a recognised international authority in this field of research.

Research results, published by Lead scientist, Professor Clare Grey of **Cambridge University** in the prestigious peer-reviewed journal Science, claims a breakthrough in the problem with current Lithium-air technology.

University of California. performance capacitors using Graphene. **University of Central Florida** Scientists have created a super capacitor battery that works like new even after being recharged 30,000 times

The Chinese company **BYD** operates a battery bank in Hong Kong with 40 MWh capacity and 20 MW

China Academy for Space Technology announced their own plans to Develop Solar Power and showcased their road map to.

The May 2014 **IEEE Spectrum magazine** carried a lengthy article titled "It's Always Sunny in Space" by Japanese Space Scientist, Dr. Susumu Sasaki.

A Publication from Dr **S. Fred Singer** Respected scientist and climate change rejecter.

Batteries News -- ScienceDaily

https://www.sciencedaily.com / New Battery research promises to could triple the range of electric vehicles --

https://www.sciencedaily.com Beyond lithium — the search for a better battery   -   FinancialTimes *https://www.ft.com/content/* Batteries - Latest research and news | Nature

https://www.nature.com Future batteries, coming

Best and worst electric cars 2018 - What Car?

Further Reading

Fact and Figures of the Earth	Longman
Scientific Lives. John Aubrey.	Hesperes
Why Does E=mc'. 2 Brian Cox	Da Capo Press
16 Bit Microprocessor Handbook.T.Raven. Newnes	
Genius of Britain Robert Uhlig.	Collins
The Next Fifty Years. John Brockman.	Phoenix
The Human Brain. Susan Greenfield.	TED Smart
The Rise of the Robots. Martin Ford.	One World
A Raven's Flight. T. Raven.	
The Book of Firsts. Steve Fossett.	Casswl llustrated
E&T . Engineering and Technology	Magazine.

Your Notes

Your Notes